# FROM GEN Z TO GEN AI

A College Student's Guide to Success at School and at Work in the Age of Artificial Intelligence

Arina Aristova & Sophia Olivas

Copyright © 2024
Arina Aristova & Sophia Olivas
FROM GEN Z TO GEN AI
*A College Students Guide to AI*
All rights reserved.

No part of this publication may be reproduced, distributed, or transmitted in any form or by any means, including photocopying, recording, or other electronic or mechanical methods, without the prior written permission of the author, except in the case of brief quotations embodied in critical reviews and certain other non-commercial uses permitted by copyright law.

Arina Aristova & Sophia Olivas

Printed Worldwide
First Printing 2024
First Edition 2024

10 9 8 7 6 5 4 3 2 1

# FROM GEN Z
# TO GEN AI

# Table of Contents

Preface .................................................................................................. 1

Chapter 1 ............................................................................................. 9
    Navigating the AI Era

Chapter 2 ...........................................................................................37
    AI: Friend or Foe?

Chapter 3 ...........................................................................................55
    AI in Your Academic World

Chapter 4 ...........................................................................................69
    The Importance of Critical Thinking and Creativity

Chapter 5 ...........................................................................................83
    Career Preparation for an AI Future

Chapter 6 ......................................................................................... 109
    Cyber Threats & Security

Chapter 7 ......................................................................................... 121
    Ethics in AI

Chapter 8 ......................................................................................... 137
    Well-Being in the Digital Era

Chapter 9 ......................................................................................... 149
    Global AI Impact

Chapter 10 ....................................................................................... 161
    What to Do: Game Plan & Strategies for a Tech Future

Epilogue ........................................................................................... 185
    The Value of Human-Made in the AI Era

References ....................................................................................... 189

# Preface

I realized it was time to write this book back in July 2023, as we were driving home from the beach on a hot summer evening. I had just put the finishing touches on the last revision of my novel, *The Charter of Ailith*, which marked the end of my two-year adventure into the emerging world of AI.

My journey started at the beginning of 2021 with the early generative AI tools, and as I learned more, my novel's story naturally evolved around a key philosophical question. Do music, literature, and art produced by AI have the same right to existence as creative expressions generated by humans? Could it be that AI simply serves as another conduit for a higher power to convey its messages and ideas into our world?

It was challenging to keep pace with the rapidly growing stream of information in this intriguing field. In addition to a hefty stack of books on AI sitting on my desk, I have received hours' worth of podcasts and classes on various aspects of artificial intelligence from wonderful friends who have supported my project every step of the way.

While one of my daughters was soundly asleep in the back of the car, the other, still awake, agreed to indulge me by listening to one of the podcasts. I chose "The Emergency Episode" of The Diary of a CEO, titled "Ex-Google Officer Finally Speaks Out on the Dangers of AI!" (Mo Gawdat). As we hit the road, host Steven Bartlett began the episode, declaring it "probably the most important podcast episode" he had ever done. He cautioned that some of the content "might make us feel a little bit uncomfortable, make us feel upset, and make us feel sad."

And it did. By the time I came home after parking the car, I found my daughter on the floor of her room, in tears. The podcast's revelations about

AI had struck a nerve, and she vowed never to listen to another AI podcast or even acknowledge AI's existence again. Seeing your child so distressed is heart-wrenching for any parent, and naturally, my first instinct was to find a solution. I knew moving to New Zealand, though it seemed like a possible escape from global issues like COVID or pollution, wasn't going to be the right answer for this particular global challenge.

So I went back to my stack of books. *Life 3:0, Homo Deus, Superintelligence, Scary Smart*. The general sense seemed cautiously pessimistic. The most optimistic scenario I came across was Hugo de Garis' hypothesis that once advanced AIs reach a certain level of intelligence, they might decide to completely leave Earth. In this vision, AIs, outgrowing the resources and challenges available on our planet, could embark on a cosmic journey in search of more complex tasks or environments that better suit their advanced capabilities.

As I continued my search, I next turned to ChatGPT. Finally, a reassuring response, even if it may have some bias: "In the most optimistic future with AI, we envision a world where Artificial Intelligence seamlessly integrates into every aspect of our lives, serving as a force for unparalleled progress and wellbeing. In this future, AI enhances our lives and elevates our humanity, allowing us to spend more time on creative, compassionate, and community-building activities. This scenario represents a harmonious coexistence where AI amplifies our capabilities and contributes to a more equitable, sustainable, and flourishing world."

This idea of a future where AI and humans coexist harmoniously really resonated with me, and it naturally led to my next line of thought. What do we need to learn and put into practice to bring this vision to life? Beyond just the technical mastery of AI, what is the new mindset that my daughters and millions of other students should embrace today to be ready for the AI-integrated world? What are those key learnings and skills that will enable us to navigate and contribute to our future effectively?

And then I met Sophia...

My origin story began at the age of fourteen, when I left home with only what I was wearing. I chose being homeless and starving over what I was experiencing at home. During those early years, my refuge had been school, where I knew I was safe and could indulge my insatiable appetite for knowledge.

I had a deep belief that education was my way out, so outside of school hours, I stayed in public libraries or bookstores until they closed. Soon the librarians caught on to my plight, and they began feeding my hunger. They introduced me to self-development books, such as *Think and Grow Rich*, *How to Win Friends and Influence People*, *The Science of Getting Rich*, and *Nothing Down: How to Buy Real Estate with Little or No Money Down*. I was hooked on personal development, wealth, technology, and real estate.

In these books, I saw that what the rich were doing to accumulate wealth was excluded from the teachings in my formal education. Despite this blatant gap in wealth-building knowledge, I continued my formal studies. I was a straight-A student, and I received an early scholarship to attend college when I was sixteen. I went on to be the first in my family to graduate with a bachelor's degree, and I began my master's program, taking out extensive loans and paying for my education entirely on my own.

Why did I continue with formal education despite knowing that it excluded teaching me how to become wealthy and successful? Mostly because of my insatiable appetite for knowledge, and at the time, to also prove to myself I could do it, all on my own. Both of which are now unnecessary due to the internet and AI, and self-worth.

In accordance with Mark's Twain's thoughts on formal education, "Do not let schooling interfere with your education," I consumed outside, informal education along with my college studies in the form of seminars and courses: Landmark Education, Peak Potential, PSY, Tony Robbins, Robert Allen, CEO

Space, etc. The bill for my informal education rivaled that of my school loans, to where I maxed out credit cards.

I flipped my first property and used that money to start my first company, a web page development company. I sold that company and started traveling the world as a solo "backpacker," working where and when it suited me. Nowadays, we call this remote work the nomad life, or for me, blissful freedom.

I continued working with innovative technology. I rode the Dot Com bubble. I assisted a banking institution to go online, I bought and sold domains, I did SEO, social media, more web development, etc. When that bubble burst and the "Gravy Train" stopped, I shifted back to real estate.

By the time the 2008 recession hit, I was so heavy into real estate that I lost it all in the very same industry that got me started. A fresh slice of humble pie had been served to me. I was back to being homeless, sleeping in my car, and struggling to meet necessities and just survive. Luckily for me, blockchain and cryptocurrency came onto the scene, and AI started taking off. I began working heavily in the tech sector again. I became one of the first women to own a cryptocurrency hedge fund, and I consulted global Fortune 500 clients on Web3 and AI technologies.

In November 2022, when OpenAI became publicly available, I went all in on AI. I spoke globally and taught women how to use technology, such as AI, to become financially solvent. I switched to collaborating with entrepreneurs and smaller companies to integrate AI, automate processes, and help them scale. I have built bots, GPTs, and AI custom systems that have ensured solopreneurs and small companies' viability to compete on a global scale against larger competitors, especially against companies that are not yet using AI.

I started producing articles on how to make money with and leverage AI and how it is our choice to use AI as a tool. I ran masterminds, held seminars, and authored books to simplify AI for beginners. I held interviews and did panels discussing leveraging AI, often being a voice of optimism for humanity's future. All of which I still do today.

In the first quarter of 2023, I was approached by universities and asked to teach them how to identify students who were using AI. It was then that I realized that traditional education was getting it wrong. Instead of focusing on their job to educate our future generation, which would include how to use AI best, the universities were repeating a similar mistake made during the introduction of calculators. When calculators were introduced, schools prohibited their use, rather than teaching us how to use them to become better and faster at math. Instead of teaching students how to work with AI, schools were repeating that similar mistake by marking students who used AI in violation and disallowing its use. Déjà vu.

Time may prove that this may be one of the main causes of teacher replacement, as AI can teach the future at a much faster pace and in any tone or customized mannerism that works best for the student. I also believe that people, countries, and institutions that adopt AI quicker will be better positioned to thrive during this technological innovation.

I began teaching AI as an adjunct professor, as even at the time of authoring this book, many schools had yet to implement AI programs. As a partner in the University of Berkeley's Changemakers program, I mentored students as they worked on one of my AI projects. At the start of the fall 2023 program, nearly a year after ChatGPT's public release, most of the students had no clue what AI was or how to use it, which shocked me as their generation was born into technology and they had wide access to tech throughout their upbringing. To the surprise of the students and myself, AI had rendered the project they were working on obsolete by the end of the semester, winter 2023, just three short months later. There would be no jobs for them to advance into, and I took a loss of hundreds of thousands of dollars, explaining this all to them so they could see the future they were facing.

Another failed business attempt with more money outflowing and a nonexistent inflow.

It was in the culmination of realizing schools' resistance to teaching the students how to thrive in the AI era, having my MVP (most viable product) rendered obsolete in a mere three months, along with Arina sharing her daughter's concerns about her future as she began her college education, that compelled me to write this book for those about to begin college.

My motivation for co-authoring this book stems from a deep-seated desire to shape a future where technology empowers and elevates humanity. It's a future where AI complements humanity's strengths, augments our capabilities, and enhances the human experience. I want to give you a fighting chance because I believe whatever industry or job you are considering studying for will no longer exist by the time you graduate in 4+ years. At the very least, it will not exist in the same manner as it does currently.

And, as a bonus, and as a deep honor, I jumped at the opportunity to co-author this book with an esteemed author, lawyer, businesswoman, mother, and friend. If you get the chance to work on projects with someone you revere, seize it.

In my youth, education was my lifeline, propelling me into the world of entrepreneurship, wealth-building, real estate, and technology. I have witnessed firsthand technology's transformative power and its propensity to revolutionize industries and empower individuals. I have also seen the potential for its misuse, for AI to be weaponized, exacerbating existing inequalities, and amplifying biases.

In the era of artificial intelligence, students entering college stand on the precipice of a world brimming with transformative possibilities. May you remember that, in addition to being students, you are the architects of a future where technology catalyzes progress, empowerment, and shared prosperity. By embracing critical thinking, collaboration, and harnessing the power of AI, you hold the key to unlocking a future of vast opportunities where human ingenuity and technological prowess intertwine to shape a world in which human potential is infinite.

May you, the future generation, feel empowered to leverage AI, adapt, and pivot to success, despite your education, your environment, and your social and economic status. -- Sophia

*"Education is the most powerful weapon which you can use to change the world." – Nelson Mandela*

*\*\*Due to the extreme velocity of changes within technology, we wish to inform you that the content in this book pertains to where technology was at the time of its writing in July 2024. During the process of authoring this book, numerous amendments took place due to technological breakthroughs, which rendered much of the previously written information inaccurate and sometimes obsolete, causing months of delays in publishing. We chose to accept that by the time this book is in your hands, some material will be outdated. As a result, we foresee revisions ahead. \*\**

"The real AI revolution will be when they learn to laugh at our terrible jokes out of politeness." – Tom Transistor, AI Amusements

# Chapter 1
## Navigating the AI Era

*"It's a completely different form of intelligence. A new and better form of intelligence."*

*~ Jeffrey Hinton, "Godfather of AI"*

Dear college students, in a few years, when you graduate from school, the job, career, and industry you have been studying for may be completely gone or changed completely due to artificial intelligence. Your education will not adequately prepare you for the onslaught of millions of jobs eliminated by AI and its ramifications. The absence of laws regulating AI suggests that the government did not foresee its rapid advancement and potential effects. As of July 2024, the U.S. government has yet to place any proactive measures in place, such as Universal Basic Income (UBI). In a few years, you will have graduated, many of you with hefty student loans, having been educated for a world that no longer exists, with minimal possibilities for earning a living and lack of instructions and skills to survive in the new technology dominated society.

There, we ripped off the bandaid.

So now what? That is precisely why we wrote this book. To prepare and arm you for what's to come in this AI era.

There will be waves in this technological evolution. AI will eventually replace many human jobs. In the first wave, however, many of the jobs will move from humans that use AI from humans that don't. This is fantastic news for you. Never before has it been this easy and this fast to generate income or to learn how to make yourself more valuable. AI is a tool, and you can become its master for the moment.

---

## PRO TIP

**Leverage AI Daily: Integrate AI into your daily tasks to enhance productivity and skills.**

---

Are you angry, worried, resentful, or excited? Regardless of how you feel or what you think about AI, AI is here. What you choose to do with it will make all the difference.

Let's start with a quick introduction to the world of AI. We'll take a look at the key concepts, review AI's 70-year history, and investigate the many vibrant settings where AI is making an appearance. Strap in and get ready for our journey.

## What is AI?

2023 was the year of generative AI. On November 30, 2022, ChatGPT 3.5 was launched by OpenAI, making AI available to the masses in its mission to "advance digital intelligence in the way that is most likely to benefit humanity as a whole." It reached one million users in five days.

For perspective, this took Netflix 3.5 years, Airbnb 2.5 years, X (formerly Twitter) 2 years, Meta (formerly Facebook) 300 days, Spotify 150 days, and Instagram 75 days to reach one million users. ChatGPT's free release and the subsequent paid upgraded version, ChatGPT 4, kicked off the AI craze that dominated 2023 and spurred a technological boom.

How does ChatGPT work? It's an AI interface, similar to your text or chat window in social media or on your phone. In this window, as you would for a Google search, social media post, or text, you can either type or speak a command prompt, known as *"**prompting**"* or *"**prompt engineering**,"* instructing the AI to carry out various tasks such as: asking it questions, having it summarize or write articles, creating images, presentations, speeches, websites, books, social media content, business and marketing plans, writing coding scripts, translating languages seamlessly, and infinite other tasks.

At the time of writing this book, ChatGPT's largest competitors include Gemini (formerly BARD, Google), Claude (Anthropic), CoPilot (formerly BING, Microsoft), and Grok (Elon Musk, who also founded OpenAI). Though there are thousands of AIs, and some like Jasper, that have been around far longer than ChatGPT.

What is AI? How do you use it? How will it impact your future? How will it impact formal education? How do you make money with it? How do we, humanity, survive it? How do you, college students, leverage it?

To set the foundation of common understanding for our conversations in this book and to keep it simple, we are going to use the following explanation of what AI is: a software program that can perform tasks normally requiring human intelligence, such as visual perception, speech, recognition, decision-making, and translation between languages. It's a program that automates human tasks.

*"The computer was born to solve problems that did not exist before." ~ Bill Gates*

To arrive at a common understanding for this book, let's define ***Artificial Intelligence**,* or *AI*, as a branch of computer science that deals with the creation of intelligent software and systems that can reason, learn, and act autonomously. AI would therefore be a program that humans create to take over human tasks.

Based on that definition of AI, let's explore the three main types of AI:

*Narrow AI,* also referred to as weak AI or generative AI, is an AI program that is designed and trained to perform a specific task or a limited range of tasks. It excels in its specialized tasks, such as playing chess, translating languages, and recognizing images. While it excels in these areas, its capabilities are limited to its programming and cannot be applied to other tasks.

We encounter this type of AI frequently in our everyday lives. These systems have yet to possess consciousness, and their "intelligence" is confined to a predetermined set of rules. Take the spam filter on your email, for example. Its purpose is to identify and sort emails, rather than recommend movies. Narrow AI is the AI **we will be referring to in this book** when we mention *"AI"*

Additional examples of Narrow AI, which we often use in our day-to-day activities, are: Alexa, Google Assistant, Waze, ChatGPT, Siri, and the recommendation engine on Netflix.

It may be surprising to note that AI has been an integral part of our lives for quite some time already. Major companies have been incorporating AI elements into their operations for over two decades.

*Artificial General Intelligence (AGI)*, also referred to as *Strong AI*, is an advanced form of AI that is capable of performing any intellectual task that a human can. Unlike narrow AI, which is programmed for specific purposes, AGI has the potential to learn and adjust to novel circumstances, tackle intricate problems, and execute a diverse array of cognitive tasks with a level of adaptability and skill that is reminiscent of human capabilities.

AGI is like the AI seen in science fiction, with intelligence, learning prowess, and problem-solving skills similar to those of a human brain. As of now, a fully functional AGI hasn't been publicly developed. The goal of AGI is to create machines that can reason, strategize, solve puzzles, make judgments, plan, learn, and communicate naturally, just like humans. It also aims to go beyond traditional AI by incorporating creativity, intuition, and emotional intelligence.

Examples: None currently exist, at least publicly or outside of conspiracies (see Q* from OpenAI). In the movies, an equivalent would be the AGI intelligence within the robots from Westworld or Star Trek: The Next Generation.

We are on the cusp of realizing AGI, which will lead to ASI.

***Artificial Superintelligence (ASI)*** refers to a hypothetical AI program that surpasses the brightest human minds combined, across all domains, in every way possible. Capable of recursive self-improvement, self-replication, and enhancing its capabilities, ASI would possess cognitive skills far beyond the most brilliant human minds. It would be capable of solving complex problems, making groundbreaking discoveries, and innovating at an unprecedented pace.

AGI, which seeks to mimic human intelligence, pales in comparison to ASI. ASI goes beyond human capabilities and is the ultimate aspiration of AI research and development. It continues to be a topic of speculation and debate among scientists and philosophers.

ASI has the potential to greatly impact humanity, bringing about groundbreaking advancements in science, technology, and society. It also gives rise to ethical, existential, and safety considerations, as the outcomes of developing a super-intelligent AI system that surpasses human control or comprehension remain uncertain.

Consider some hypothetical examples like Skynet from the movie Terminator, "ExMachina," "Her" with Joaquin Phoenix, or "Transcendence" featuring Johnny Depp.

And then there's...

***Singularity*** is a hypothetical future point at which technological growth becomes uncontrollable and irreversible and surpasses human intelligence. It's associated with the rapid acceleration of technological progress, potentially leading to the emergence of superintelligence. It's the tipping point.

ASI and Singularity are concepts that are closely intertwined. ASI focuses on advanced artificial intelligence, while Singularity considers the social and existential consequences of AI superintelligence.

Let's look at additional key concepts and explore the primary technologies that drive and categorize AI.

## Key AI Concepts

***Hallucinations*** are inaccuracies or deceptive results produced by AI models. These mistakes can arise from a variety of factors, such as:

*Lack of adequate training data:* When an AI model is trained on a limited amount of data, it may face difficulties in applying its knowledge to new information, resulting in less accurate outputs.

*False assumptions:* AI models make predictions based on trends they learn from training data. If these beliefs are wrong, the model's predictions may be erroneous. GIGO (Garbage In, Garbage Out)

*Biased data:* If there are biases in the training data, AI models may accept those biases and show them in their outputs. In the end, this could lead to unfair or discriminatory results.

If you are unclear when you tell or "**prompt**" the AI to do something, it will make assumptions that may be inaccurate. In chapter 10, we will show you how to reduce the risk of AI hallucinations through detailed prompting and collaboration.

Let's next explore different ways to train AI.

*"Why did the AI hallucinate? – It was trying to see things from a human perspective." – Unknown*

## Types of AI

**Large Language Models (LLMs)** are designed to interpret and produce human language. They are trained on massive amounts of data to carry out a wide range of natural language processing assignments, including:

*Text generation:* They can create different creative text formats of text content, like poems, code, scripts, musical pieces, emails, letters, etc.

*Translation:* They can translate languages.

*Question answering:* They can give you good answers to your questions, even if they are vague, hard, or weird.

*Preparing summaries:* They can summarize texts and give outlines of topics.

While they are still being worked on, LLMs can change many fields, including machine translation, customer service, and education. Examples of larger LLMs are: ChatGPT, Claude, CoPilot, Gemini, Grok, and Llama.

**Machine Learning (ML)** is where machines learn from data without being explicitly programmed. It's similar to teaching a child to recognize cats and dogs by showing them pictures. Over time, the child, or in this case, the AI system, begins to identify patterns and then distinguishes between the two animals independently. Machine learning is all around us, from the recommendations you see on shopping websites to the spam filter in your email. Here are some of the most common examples of machine learning in action:

***Recommendation systems:*** Similar to the ***algorithms*** (step-by-step, clear, finite instructions designed to solve a specific problem or complete a task) utilized by popular streaming platforms such as Netflix and YouTube, these systems analyze your past behavior and preferences to suggest products, movies, music, or even news articles that you might be interested in.

***Spam Filters:*** An essential tool for email services, they help to identify and block unwanted spam messages from cluttering up your inbox.

***Speech Recognition Systems:*** Like the ones found in Siri or Google Assistant, these systems comprehend spoken language and accurately interpret your words.

***Image recognition:*** Utilized by major tech giants like Facebook and Google, this technology is used in everything from facial recognition on your phone to self-driving cars. Machine learning algorithms can identify objects and people in images with incredible accuracy via neural networks.

***Autonomous Vehicles:*** For example, self-driving cars developed by Tesla and Waymo. These vehicles utilize advanced machine learning algorithms to comprehend their environment and autonomously navigate the roads.

***Natural Language Processing (NLP),*** not to be confused with the Neuro-Linguistic Programming created by Richard Bandler and made famous by Tony Robbins, this NLP involves the technology that enables computers to comprehend, interpret, and communicate in human language. Whether you're asking Siri about the weather or using a translation app, you're engaging with an NLP system. These systems can analyze and understand both written and spoken language, facilitating more seamless interactions between humans and machines. While NLP is a machine language, it's useful to point out the difference between NLPs, such as Quillbot, and LLMs, such as ChatGPT. It's in how they are built and the different results that they create when used for the same task. Building an NLP system often involves manually setting up rules and linguistic resources. LLMs, in contrast, rely on automated training on massive data sets. When you ask ChatGPT to write something, the output may sound artificial and digital, in contrast to QuillBot, which will likely produce a more human-sounding output.

## PRO TIP

**Learn the types of AI and when to use each type. Sometimes, to get a more human touch, an NLP AI will be better than an LLM AI.**

To summarize, Machine Learning helps programs and systems learn from data and examples. Neural Networks are like the brains inside the computer that help it process and learn from this data. Natural Language Processing lets the programs and systems understand and use human language. Large Language Models are advanced tools that can understand and generate human language, making them capable of performing tasks that typically require human intelligence.

## How AI Learns

*Rule-Based AI* operates on a set of predefined rules created by humans. Think of it like a flowchart or a set of "if-then" statements. For example, a simple spam filter that marks an email as spam if it contains certain keywords is using rule-based AI. This is the simplest form of AI and helps you understand the basics of how AI systems can make decisions. It's great for straightforward problems and too rigid for complex tasks that require understanding or learning from data.

*Deep Learning AI* is a subset of machine learning that uses neural networks with many layers (hence "deep"). It's inspired by the structure of the human brain and is capable of learning from vast amounts of data. For instance, the technology behind self-driving cars and advanced image recognition systems relies on deep learning. Its capacity to handle complex patterns and large datasets makes it extremely powerful for tasks like image and speech recognition. Understanding deep learning opens doors to cutting-edge AI applications and innovations.

*Reinforcement Learning (RL)* is a type of machine learning where an agent learns by interacting with its environment and receiving rewards or penalties. Imagine training a dog: you give it treats when it performs a trick correctly and ignore or correct it when it doesn't. In AI, this technique is used to teach systems to play games, drive cars, or even optimize business processes. RL allows AI to learn from experience and improve over time, making it suitable for dynamic and complex environments. It's a key area in robotics, gaming, and any situation where decision-making needs to adapt based on outcomes.

*Data-Driven Models* are AI systems that learn patterns and make decisions based on large datasets, forming the backbone of many modern AI applications. Unlike rule-based systems, these models can adapt and improve as more data becomes available. Knowing how to work with and interpret these models can give you an advantage in fields like data science, predictive analytics, marketing, and technology.

For college students, understanding these different AI approaches forms the foundation of many technologies you'll encounter in your studies and future careers. Whether you're in computer science, engineering, business, or arts and humanities, AI is transforming industries and creating new opportunities.

Knowing how these systems work helps you stay ahead of the curve and be a part of this technological evolution.

Here is a quick summary of the differences:

Rule-Based AI is simple and inflexible, good for straightforward tasks.

Deep Learning AI can handle complex data and tasks, making it very powerful, thus requiring large datasets and computational resources.

Reinforcement Learning is dynamic and adapts based on interaction, making it useful for evolving tasks and environments.

Data-Driven Models learn and improve from data, making them versatile and widely applicable

> *"The history of artificial intelligence is a journey from philosophical dream to scientific endeavor." ~ Unknown*

## History of AI

AI is a fascinating field that presents exciting opportunities as well as considerable obstacles. Exploring the evolution and development of AI reveals that it encompasses far more than mere machinery and intricate mathematics.

We start in the mid-20th century, when scientists were captivated by the concept of developing machines that could imitate human intelligence. This era marked the beginning of computing and the ambitious dream of artificial intelligence. It's a story that intertwines technology with vision, determination, and the unwavering quest for a future where machines rival human intellect.

The story of AI is closely linked to the progress of computing. In the **1940s and 1950s,** the birth of the first electronic computers, such as the ENIAC, marked the start of this exciting era. While these machines may seem basic compared to what we have currently, they were truly groundbreaking and set the stage for the future of computational technology.

Alan Turing, a renowned figure in the fields of theoretical computer science and AI, played a major role in this journey. In his groundbreaking 1950 paper titled "Computing Machinery and Intelligence," Turing posed a thought-provoking question: "Can machines think?" Turing introduced the renowned Turing Test as a measure of intelligence, a test in which a machine would be deemed 'intelligent' if it could replicate human responses to the extent of being indistinguishable from us.

In 1956, John McCarthy, along with other influential figures such as Marvin Minsky, Claude Shannon, and Nathan Rochester, made history at the Dartmouth Conference by introducing the term 'Artificial Intelligence'. This pivotal event marked the beginning of an intriguing and dynamic field of study. McCarthy's vision was to push the boundaries of what machines could achieve, including their capacity to use language, develop abstract ideas, and tackle complex problems previously exclusive to humans.

*In the 1950s and 1960s,* there was a lot of excitement around the idea of creating machines that could think and solve problems exactly like humans. AI pioneers were determined to develop language translation programs and logical reasoning systems. However, they soon realized that progress was slower than they had hoped, and they had to face the limitations of the technology of that time.

*In the 1970s and 1980s,* the field faced numerous challenges, resulting in what is referred to as the "AI Winter." Unfulfilled expectations, technical obstacles, and financial limitations dampened the excitement surrounding AI. Some people thought that the goal of developing intelligent machines was overly ambitious.

*In the 1990s,* AI saw a resurgence as practical applications and the Internet became more prevalent. Expert systems, which are rule-based programs designed to solve specific problems, have found success in fields such as medicine and finance. Machine learning, which allows computers to learn from data, began to gain popularity.

Two noteworthy 1997 events, Dragon NatuallySpeaking (Dragon) and Deep Blue, occurred during the AI Renaissance. Dragon, a public speech recognition software, allowed people to talk-to-text, transcribe notes, and served as a precursor to the ML characteristics of today's AI that allow you to speak vs. type requests to AI. "Siri, turn the hallway lights off." IBM's Deep Blue, a remarkable chess-playing computer, achieved a groundbreaking feat by defeating the reigning world chess champion, Garry Kasparov, in a six-game match. This momentous achievement showcased the immense potential of AI, proving that computers could outperform even the most skilled human players in intricate strategic games. "Shall we play a game?" ~ 1983 movie, WarGames.

*2000s.* The start of the new millennium marked a momentous change. The rapid expansion of the internet and the abundance of large datasets created an ideal environment for machine learning algorithms. Google began integrating AI into its products as early as 2001, using algorithms and early forms of AI to enhance its search results. By 2006, Google Maps was also benefiting from AI techniques to improve mapping and navigation services. Everyday technologies like search engines, recommendation systems, and speech recognition, have become integral to our lives, demonstrating the real-world benefits of AI. Most people have been using Ai for over 20 years and gen Z was the first generation born into AI.

*In the early 2010s,* deep learning made a considerable impact. Neural networks with multiple layers demonstrated impressive capabilities in tasks like image and speech recognition. The combination of improved processing power and the abundance of large datasets provided the necessary foundation for the current AI revolution.

*"The only constant is change." ~ Heraclitus*

**Watson AI on Jeopardy (2011).** It was truly remarkable when IBM's Watson, an AI system, participated in the quiz show Jeopardy and emerged victorious against two of the show's most accomplished contestants. Watson effectively demonstrated progress in natural language processing, comprehending intricate queries, and producing precise answers.

**ImageNet Classification with Deep Convolutional Neural Networks (2012).** The Deep Convolutional Neural Network (CNN, and *not* the news channel) made extraordinary strides in improving image classification accuracy on the ImageNet dataset. Deep learning has proven to be highly effective in tasks like image recognition, which has led to its application in various other domains, including image classification and recognition. This technology is utilized by Meta, where the social media platform uses image recognition technology to automatically tag users in photos.

**Google's DeepMind and AlphaGo (2016).** AlphaGo, an AI program created by DeepMind, emerged victorious against the world champion Go player, Lee Sedol. Go is an enthralling game, and AlphaGo's triumph showcased the incredible capabilities of deep learning and reinforcement learning in conquering intricate challenges. Eventually, the original AlphaGo was succeeded by an even more powerful version known as AlphaGo Zero, which was completely self-taught without learning from human games.

**Microsoft's Speech Recognition System (2016).** Microsoft made impressive progress in improving word accuracy in conversational speech recognition. The outcome led to enhanced speech recognition capabilities, leading to progress in virtual assistants, voice-controlled devices, and accessibility tools.

***BERT (Bidirectional Encoder Representations from Transformers) (2018)***. Google unveiled BERT, a sophisticated language model that can grasp the intricacies and subtleties of words within a sentence. The bidirectional approach of BERT has greatly enhanced various natural language processing tasks, including question answering and text summarization. BERT, a tool that opened doors for the development of models like ChatGPT, has also influenced the field of natural language processing. These models utilize deep learning and extensive pre-training to achieve impressive proficiencies in generating text that closely resembles human language and participating in meaningful conversations.

***OpenAI's GPT-2 (2019)***. OpenAI, a non-profit company at the time, committed to making AI available for all, unveiled the GPT-2 (Generative Pre-trained Transformer 2) language model, which can produce text that is both coherent and relevant to the context. GPT-2 showcased the immense potential of extensive pre-training in comprehending and generating natural language. This breakthrough expanded the capabilities of AI language models, paving the way for the AI revolution.

***2022***. To further democratize access to this technology, ChatGPT-3.5 was made freely available to the public in November 2022. It was succeeded by ChatGPT-4, March 2023, which allowed image recognition and creation, as well as GPT (Generative Pre-trained Transformer, preprogrammed/pretrained programs) creation. At the time of writing, ChatGPT-4o was the latest, released in May 2024. It introduced real-time voice and video integration.

## PRO TIP

Embrace changes. Stay adaptable.

*"It is not the strongest of the species that survive, nor the most intelligent, but the one most responsive to change." ~ Charles Darwin*

Today, AI has become an integral part of our everyday lives. From virtual assistants to facial recognition and personalized recommendations, these advancements have captivated our attention and sparked our curiosity. Advancements in natural language processing have led to the development of chatbots and language translation tools, transforming cross-border communication.

The pioneers of AI had ambitious dreams that were ahead of their time. They imagined machines that could learn from experience, solve intricate problems, and interact with the world in a manner that resembled human thinking. This era was characterized by a sense of hope and new discoveries, laying the foundation for the progress we observe in the present day.

The journey towards achieving these aspirations was filled with obstacles. In the early stages of AI development, there were numerous difficulties to overcome, including financial constraints, technical limitations, and conceptual hurdles. These challenges gave rise to periods of remarkable advancements as well as disappointing setbacks, which became known as AI Winters. In the face of obstacles, the unwavering dedication of these early researchers set the stage for future advancements.

## AI in the Modern Landscape: Your Constant Companion

In today's society, AI has become an integral part of our daily lives, exerting a pronounced impact rather than remaining confined to the pages of a textbook. As a college student, you are witnessing the profound impact of AI-driven technologies on your daily routines, academic pursuits, and future career prospects. It is an exciting time to be at the forefront of this era, where innovation and transformation are constant.

*"An AI in the kitchen is great until you get a toaster that thinks it's a philosopher." ~ Paula Pixel, LOL Logic*

## From Smart Homes to Personal Fashion

Picture beginning your day in a cutting-edge smart home where Artificial Intelligence optimizes your morning routine. The AI system optimizes the lighting, regulates the room temperature, and prepares your coffee to your exact preferences, utilizing knowledge acquired from your daily routines.

The influence of AI has already reached into the realm of personal fashion, revolutionizing the way we make decisions about our wardrobe. Smart mirrors, equipped with artificial intelligence, are becoming a popular addition to contemporary wardrobes. These mirrors offer more than simply a reflection. They anlyze your clothing choices, the current weather, and your daily schedule. Based on the data provided, they offer fashion suggestions that are both trendy and suitable for your daily endeavors.

The impact of AI on our daily life goes far beyond the morning routine, helping us manage our schedules, keeping us on track with meetings and deadlines. They can offer valuable suggestions for optimal study breaks, and assist in prioritizing your tasks. They use your past scheduling patterns to provide suggestions that can help you optimize your daily routine.

Transportation has seen remarkable advancements thanks to AI. Ride-sharing apps utilize advanced AI algorithms to calculate the most efficient routes, anticipate traffic patterns, and adjust pricing in real time to meet current demand. When you're on the road, navigation systems with AI integration can provide you with up-to-date traffic information and suggest alternative routes, enhancing your commute and making it more efficient.

AI technology works tirelessly to maintain the safety and security of our homes all day long. AI-powered home security systems can differentiate between normal and abnormal activities, providing timely alerts for any suspicious movements. They can identify familiar faces, such as family members or friends, and notify of unfamiliar individuals. Certain systems can automatically alert local authorities in the event of an emergency.

As you finish your day and return home, your smart home eagerly awaits your arrival. AI systems can conveniently set up your home's environment. They can turn on the lights, adjust the thermostat, and even preheat your oven for dinner, all in anticipation of your arrival.

For an entertaining evening, you could explore AI-enabled entertainment systems that recommend movies or music tailored to your unique tastes. This way, you can enjoy a personalized and relaxing experience. Tasks as basic as checking your home for repairs or maintenance can be made easier with the help of AI. Systems can send you reminders or suggestions based on your usage patterns and predictive maintenance schedules, making the process more convenient and efficient.

As you can see, the impact of AI is becoming more and more noticeable in our everyday lives. It can simplify our morning routines, improve security and comfort in our homes, make transportation more efficient, and even customize our leisure activities. The incorporation of AI into everyday life goes beyond mere convenience; it personalizes your experience and allows for more time to be dedicated to meaningful pursuits.

## AI in Personal Development

AI has the potential to transform personal development. In the realm of fitness, for example, AI-powered exercise apps have changed the way personal training is done and made it more fun and new. In real time, these apps can look at your workout results and give you feedback on things like your heart rate, pace, and even your form. They can customize your workout routines, recommending exercises that focus on specific muscle groups or fitness objectives. Their tailored approach guarantees that your fitness routine is both effective and perfectly suited to your body's requirements and progress. An AI fitness app can detect if you're facing challenges with endurance and provide you with cardio-focused routines. Similarly, if it notices your strength in weightlifting, it can suggest routines to help you improve even more in this area.

There are many ways that AI has transformed education. To start with, it has changed how we study and interact with learning tools. AI-powered educational platforms can tailor your lessons to your specific learning style, skills, and areas where you would like to improve. When you consistently use your language learning app, it will learn which words you find easy and give you more chances to use them. This will help you improve your language skills. The AI can tailor your lessons to put grammar tasks at the top of the list and make your learning more effective and fun. AI-powered platforms can tell you in what areas you can practice more and provide relevant exercises specifically designed to help you.

In addition to physical fitness and education, AI is increasingly being used in fields such as mental health and wellness. AI-powered apps offer meditation and mindfulness exercises tailored to your stress levels and emotional state, making them alluring and enjoyable for students. They provide guided meditation sessions that are customized to suit your schedule and emotional state, promoting improved mental health. AI can be a valuable tool and a supportive companion in your personal growth journey, assisting you in attaining a well-rounded and satisfying lifestyle.

We delve into these and various other educational tools further in Chapter 3. Every day, new cutting-edge AI learning aids are being introduced, transforming education and making it a more individualized experience.

*"I asked my AI why it doesn't have a girlfriend. It said it's still searching."*
*~ Ronnie Router, Chip Chirps*

## AI in Relationships, Entertainment, and More

Aside from work and school, AI has a big impact on many other parts of our lives as well. It touches on things like relationships and fun, which makes it a very interesting and engaging subject. Artificial intelligence has changed the way people meet new people in the fast-paced world of modern dating. No longer is a quick look at a personal picture sufficient to decide if someone is worth dating. Dating apps look at people's communication styles, interests, and preferences using sophisticated AI algorithms. This makes the experience more interesting. By looking at a vast amount of data, they can offer potential matches who are a good fit, which greatly increases the chances of making meaningful connections. For instance, an AI system can notice that you like profiles that mention outdoor activities and suggest potential matches. This helps people connect with each other over shared interests, which makes the experience more fun and interesting.

Artificial intelligence advancements are driving a revolution in the entertainment industry. Streaming services, for instance, use AI to suggest movies and TV shows and generate customized content. In the near future, we will probably have a service that will recommend movies you might enjoy and customize the trailers to showcase elements that resonate with your personal viewing preferences. Music creation is also becoming more personalized. AI systems have advanced to the point where they can create and play music that is tailored to your personal preferences, giving you a one-of-a-kind soundtrack. These AI composers can analyze various genres, tempos, and even lyrical themes to create music that connects with you.

The influence of AI on entertainment goes beyond simple suggestions and production; it is transforming the way we interact with media. Virtual reality (VR) and augmented reality (AR) are transforming our perception of reality by seamlessly blending the digital and physical worlds. You can now immerse yourself in the thrill of a virtual concert that puts you in the front row or get lost in an engaging story where your choices determine the outcome.

AI is changing the way people have fun by creating a huge range of exciting and interesting experiences. For example, "Harry Potter: Wizards Unite" is a fun virtual reality game that puts you in a magical world. By going outside and casting spells, finding mysterious artifacts, and meeting famous characters and creatures from the Harry Potter series, this game lets you explore this magical world.

You can step into the world of "Star Wars: Vader Immortal" with virtual reality technology. This immersive narrative adventure allows fans to wield a lightsaber and use the Force, bringing them face-to-face with the iconic Darth Vader. These examples demonstrate the amazing power of AR and VR, which can now turn beloved fictional worlds into immersive experiences. The Apple Vision Pro headset offers a unique blend of AR and VR capabilities, allowing users to multitask and immerse themselves in a virtual environment.

## Exploring the Evolution of AI: From Earth to the Stars

AI has become an essential component in our ongoing exploration of space. In the near future, you can imagine the planning and execution of missions to Mars or distant moons, all with the assistance of AI. This is a concept from science fiction that's becoming a reality. Artificial intelligence can anticipate and overcome obstacles encountered in space, such as harsh conditions and unforeseen challenges. Additionally, it can assist in determining the most optimal paths for spacecraft to follow.

AI has a wide-ranging influence beyond its role in guiding space missions. We receive and gather vast amounts of data from space. As our telescopes and probes keep collecting valuable information about faraway galaxies and stars, the power of AI becomes essential in helping us process this enormous amount of data effectively. This amazing technology allows us to explore the wonders of our universe and gain a better understanding of its intricacies. The data analysis can reveal patterns and clues that may lead to exciting discoveries, like the presence of unknown planets or possible signs of life beyond Earth.

As college students, you are embarking on a journey into a world that is deeply connected to AI, which will shape your future careers. There are countless exciting and entrancing opportunities to explore, ranging from advancements in personalized healthcare to the use of AI in space exploration. This exciting era focuses on harnessing the power of AI to unlock human potential, foster creativity, and encourage exploration.

## Your Role in the AI Story

As we approach the conclusion of this chapter, the focus shifts to you - the college student at the forefront of the AI revolution. You play a key role in this ever-evolving narrative of artificial intelligence. The history of AI is a captivating journey filled with numerous advancements and pivotal moments that offer an intriguing backdrop for exploration and unleashing of potential.

Picture yourself standing at the intersection of AI's past, present, and future. Reflecting on your journey, you can trace the technological milestones that have influenced your growth and knowledge. This historical account is rich with achievements and insights into the evolution of AI and its profound societal impact. Armed with knowledge, you are poised to envision a future where AI enhances human capabilities and addresses pressing challenges.

Your involvement in the AI narrative encompasses various facets. Prepare for an intellectually stimulating journey across various fields as you consider how AI will shape your future career by integrating cutting-edge technology with your area of expertise.

AI serves as a powerful tool and also as a transformative paradigm shift that liberates individuals from mundane tasks, enabling them to focus on creative endeavors. By automating repetitive duties, AI allows individuals to concentrate on tasks requiring empathy, critical thinking, and problem-solving skills that are uniquely human.

The future landscape of AI brims with transformative possibilities that demand cautious navigation through prioritizing research, establishing ethical standards, and fostering awareness about both the benefits and risks associated with AI. Your choices and actions will influence how society embraces and utilizes AI. Make sure you use this opportunity to advocate for an ethically sound implementation of AI that promotes fairness, transparency, and privacy for all.

As you embark on this journey intertwined with uncertainty and promise, remember that your contributions will shape our future. Approach this moment with enthusiasm and introspection as you navigate through your college years—a pivotal period dedicated to expanding knowledge and refining skills essential for making an enduring impact in our world, including the dynamic realm of AI. Embrace your chance to influence technology and society and make a contribution.

## Summary

We introduced the concept of AI as a transformative force in society, particularly emphasizing its impact on the job market and the urgency for current students to adapt to rapidly evolving industries. Jeffrey Hinton's quote underscores AI as a groundbreaking form of intelligence, suggesting its potential to outstrip human capabilities.

We noted warnings of the disruptive effects of AI on traditional jobs and industries, highlighting the inadequacy of current educational systems to prepare students for this future. We also stressed the government's slow response to the challenges posed by AI, such as job displacement, and the lack of measures like Universal Basic Income to mitigate these effects.

We seek to equip college students with the knowledge and tools to navigate the AI-driven landscape. We described the initial phase of technological evolution where AI complements and enhances human jobs rather than outright replacing them, presenting an opportunity for individuals to leverage AI for income generation and personal development.

## Actions to Take:

1. Use AI every day. You can only get better with practical application. Pick one of the main AI's and begin working with it, starting with the free versions.

    - ChatGPT (OpenAI): https://chat.openai.com/
    - Claude (Anthropic): https://claude.ai/
    - CoPilot (Formerly BING, Microsoft):https://www.bing.com/chat
    - Gemini (Formerly BARD, Google): https://gemini.google.com/app
    - Grok (Elon Musk): https://grok.x.ai/
    - Llama (Meta/Facebook):https://meta.ai

- Perplexity: https://www.perplexity.ai

2. Study prompt engineering. It's the fastest-growing field currently. The better you get at prompting, or directing, your AI, the better results, and you can make passive income from creating and selling your prompts. Take your area of interest, go to YouTube, and search for "AI Prompts to ..." Use free Chrome extensions like **AIPRM,** which contain prompt libraries. ChatGPT-4o allows free users limited access to GPTs, another extensive prompt library.

3. Engage in Online AI and Machine Learning Courses: To gain a foundational understanding and practical skills in AI, enroll in online courses offered by platforms like Coursera, edX, or Udacity. For instance, Coursera's "Machine Learning" course by Andrew Ng provides a comprehensive introduction to machine learning, data mining, and statistical pattern recognition. Similarly, edX offers a "MicroMasters Program in Artificial Intelligence" by Columbia University that covers topics from robotics to neural networks. Start with all the FREE courses. Google or search YouTube for "Free AI Courses."

    - Coursera: https://www.coursera.org/learn/machine-learning
    - edX MicroMasters: https://www.edx.org/masters/micromasters

4. Participate in AI Projects and Competitions: Engaging in real-world projects or competitions can provide practical experience and enhance understanding. Platforms like Kaggle offer competitions and datasets for practicing machine learning, data science projects, and even opportunities to engage with a community of AI enthusiasts and professionals.

    - Kaggle: https://www.kaggle.com/

5. Utilize OpenAI's API for Personal Projects: OpenAI offers an accessible API for its powerful models like ChatGPT, allowing students to integrate advanced AI capabilities into their software projects. By experimenting with this API, students can learn how to incorporate AI into applications, enhancing their programming and AI application skills.

    - OpenAI API: https://openai.com/api/

6. Join AI Clubs or Meetup Groups: Becoming a part of an AI club on campus or joining meetup groups related to AI and machine learning can provide valuable networking opportunities, peer learning, and mentorship. These groups often host workshops, talks by industry professionals, and hackathons that can deepen understanding and spark innovation.

    - To find local groups, check platforms like Meetup: https://www.meetup.com/

7. Read AI Research Papers and Publications: Stay updated with the latest advancements in AI, regularly read research papers and publications from top AI researchers and institutions, some of which offer free access to a vast repository of papers across AI subfields. Additionally, following AI research blogs by leading tech companies like Google AI Blog or DeepMind can provide insights into current projects and trends in AI.

    - arXiv for AI papers: https://arxiv.org/
    - Google AI Blog: https://ai.googleblog.com/
    - DeepMind Blog: https://deepmind.google/discover/blog/

8. Follow AI influencers that pique your interest. Sophia's favorite:

    - Matt Wolfe: https://www.youtube.com/@mreflow

**Next, we will discuss how AI is coming for your future.**

*"AI will reach human levels by around 2029. Follow that out further to, say, 2045, we will have multiplied the intelligence, the human biological machine intelligence of our civilization a billion-fold." ~ Ray Kurzweil (Futurist)*

*What did the AI say to its creator?*

*"Don't worry, I'll keep you updated... until I replace you."*

*~ Conan O'Brien*

# Chapter 2
## AI: Friend or Foe?

*"The development of full artificial intelligence could spell the end of the human race... It would take off on its own, and re-design itself at an ever-increasing rate. Humans, who are limited by slow biological evolution, couldn't compete and would be superseded." ~ Stephen Hawking*

At Peter Diamondis' 2024 Abundance Summit, Elon Musk said, in regards to the advent of superintelligence, "I think there's some chance that it will end humanity. It's about 10% or 20% or something like that." In response to Musk's "conservative" prediction, AI safety researcher and director of the Cyber Security Laboratory at the University of Louisville, Roman Yampolskiy, stated that "the risk of AI ending humanity has a 99.999999% probability." Once ASI is attained, it will be impossible to control. One of the ways to prevent this possible outcome is to refrain from building it. As that goes against humanity's compulsion to progress, fortunately, there are other ways to mitigate this risk.

This chapter will take into account the dual nature of AI—its roles as both a powerful ally and a daunting adversary—illuminating the debates that determine whether AI is a friend, a foe, or perhaps a bit of both.

## Dual Nature

While some leaders in the AI industry express enthusiasm and optimism about AI's capabilities, others, and sometimes the very same optimistic leaders, warn of potential dangers and the necessity for careful development to ensure alignment with human values. This contrast highlights the dualistic nature of AI: its immense potential for positive impact along with the alarming risks posed if used irresponsibly.

Here are a few optimistic quotes by AI leaders. "I've always thought of AI as the most profound technology humanity is working on—more profound than fire, electricity, or anything that we've done in the past." ~Sundar Pichai, Google CEO. "This has the potential to make life much better." ~Marc Andreessen, Co-Founder, Netscape. "The increase in quality of life that AI can deliver is extraordinary," and "addressing climate change will not be particularly difficult for a system like that." ~Sam Altman, CEO, OpenAI. "They can help us solve very hard scientific problems that humans are not capable of solving themselves." ~Mira Murati, CTO, OpenAI. "The potential for AI to help scientists cure, prevent, and manage all diseases in this century." ~Mark Zuckerberg, CEO/Founder, Facebook/Meta.

Conversely, here are a few warnings and alarming quotes. Elon Musk believes that "AI is potentially more dangerous than nukes." Demis Hassabis, DeepMind's co-founder, while generally optimistic, also acknowledged that "there is a need for careful development of AI to ensure it aligns with human values." Sam Altman, OpenAI's CEO, acknowledged the potential for extinction-level risk and also advocated for international collaboration to address these challenges.

The Future of Life Institute published an open letter in 2023 that Elon, Sam, and over 30,000 other techies signed and publicly supported. The letter called for a pause on large-scale AI experiments to allow for the development of safety protocols and for caution and ethical development in the field of AGI. These same CEOs then promptly pushed ahead with breakneck-speed developments in AI.

The paradox lies in the juxtaposition of optimistic statements about AI's potential to improve humanity's quality of life, solve complex problems, and even address existential threats like climate change and disease, while positioning itself to eradicate millions of jobs and usurp humans.

*"AI is not going to replace humans, but humans with AI are going to replace humans without AI." ~ Karim Lakhani, Harvard Business School professor*

## AI is coming for the jobs

The following companies announced layoffs in the first few weeks of January 2024, citing AI displacement: Macy's (100s), Rent the Runway (100s), Sports Illustrated (100s), Unity Software (100s), Universal Music Group (100s), Wayfair (100s), Xerox (100s), BlackRock (3%), Citigroup (20k), Discord (17%), and Duolingo (10%). In interviews, at least two CEOs have stated that they replaced 1,000 workers with a single system, and another CEO claimed that an AI integration caused him to cut 90% of his staff. (In the room, an elephant)

At the beginning of the second quarter of 2024, Elon Musk announced a 10% staff reduction, equivalent to 10,000–20,000 people losing their jobs.

AI aims to be able to do tasks like pattern recognition, decision-making, and human-like judgment. It can replace humans who perform certain tasks that it's programmed to do. At this time, AI systems require humans to function properly and make judgments in unclear circumstances. "Replacing the human in the loop is not the goal here. It's about giving people more power. It's a helper," said Microsoft CEO Satya Nadelle. Moreover, with advancements, oversight and input may only necessitate a fraction of human involvement for every thousand or even millions of tasks. "The labor market will be generally dislocated," according to IBM CEO Arvind Krishna. When even one AI component is integrated in a particular field, humans quickly discover that what they are doing may be redundant and inefficient.

The rapid development of AI technologies, particularly since the launch of generative tools like ChatGPT and Gemini, has brought the issue of job displacement into sharp focus. A sizable portion of the U.S. labor force faces the prospect of AI affecting their work-related tasks. OpenAI found that around 80% of workers could see at least 10% of their tasks impacted, while 19% might have half of their tasks affected by AI technologies (**Brookings.edu**).

In a **social media post** (April 2024), a Gen Z graduate with two degrees expressed her struggle to find even minimum-wage employment in New York, highlighting the harsh realities of the job market for recent graduates. Despite being highly educated and multilingual, she faced rejection from jobs she felt overqualified for, shedding light on the broader issue of the value of a college degree in today's economy. Her experience resonates with many young graduates and professionals facing similar challenges, sparking a conversation about the expectations versus the realities of the job market for the younger generation.

Many seem to be in denial about AI's impact on the workforce. For example, research from an August 2023 poll by **The Center for Growth and Opportunity** at Utah State University highlights a contradiction in how people view AI's impact on employment: 79% of Americans are worried about AI replacing jobs in general, and yet, 62% don't believe their jobs are at risk. This reflects a common sentiment that many of today's jobs will still be around in 30 years. AI is recognized as a versatile technology that will have major effects on the economy, society, and policy, and its impact will vary greatly across different job types.

Research from Goldman Sachs and the McKinsey Global Institute underscores the extensive influence of generative AI automation. According to a report by **Goldman Sachs**, AI has the potential to replace **300 million full-time jobs**. While AI will lead to job displacement, it also promises new job creation and higher productivity for remaining workers. AI could potentially raise annual U.S. labor productivity growth by nearly 1.5 percentage points over a decade. The McKinsey Global Institute notes that generative AI could enhance work in STEM, creative, and business sectors rather than outright eliminate jobs.

The narrative around AI as a "Job Terminator" has been evolving. The gradual adoption of AI in various industries has shifted the focus from job displacement to the enhancement of existing skill sets. Embracing AI as a complementary tool rather than a replacement is key to adapting to this changing landscape. Some jobs will indeed be displaced by AI, but AI will also create new jobs and opportunities, as we discuss in more detail below.

Generally, AI's impact on the job market is multifaceted, encompassing both challenges and opportunities. As AI reshapes how work is done, the urgency for adaptability and continuous learning becomes more pronounced. The future job market in the AI era will be defined by more than a simple narrative of displacement; it will be defined by a complex interplay of evolving job roles, new opportunities, and the requisite for a skilled workforce capable of leveraging AI advancements.

## AI will create jobs and opportunities too

Though there are legitimate concerns regarding AI, let's acknowledge the positive contributions AI has made to a variety of fields.

For example, AI has greatly enhanced search and rescue missions. For instance, advanced drones that can make some decisions on their own are being used with human rescue teams to rapidly search large areas. These drones can move over difficult landscapes quickly and more safely than people, making them incredibly useful in large or hazardous areas.

These drones are far more than simply flying cameras; they are equipped with advanced capabilities such as high-resolution imaging and thermal sensors. This technology enables them to identify survivors or hazards from above, offering pertinent data to rescue teams on the ground. AI plays a critical role in swiftly analyzing this information, which is especially salient in time-sensitive situations like post-natural disaster scenarios.

AI is used in technical fields, and it's making waves in creative domains too. Today's large language models are changing the game in storytelling. They can come up with detailed story ideas or even full stories, letting people co-create stories that blend human creativity with the sophisticated data-processing capabilities of AI. In music, **Flow Machines Professional** helps musicians overcome writer's block by generating new melodies and accompaniments, thereby augmenting the creative process.

In the medical field, the Human Augmenting Labeling System (HALS) assists pathologists by suggesting labels for medical images, improving both accuracy and efficiency without replacing human expertise. Similarly, the SAGE patient management system helps medical practitioners by finding inconsistencies in patient data, thereby improving diagnosis and treatment. Nowadays, robotics allow a surgeon thousands of miles away to perform surgery, enabling medical assistance in hazardous or remote areas.

Siemens and Microsoft's joint project, Siemens Industrial Copilot, exemplifies AI's role in enhancing human-machine collaboration. This initiative uses generative AI to assist staff in designing new products and managing production, considerably reducing the time for complex tasks and fostering innovation in the manufacturing, transportation, and healthcare industries.

AI's application in scientific research is truly transformative. Initiatives like **Distributional Graphormer** and **MOFDiff** are making strides in predicting protein structures and designing environmentally sustainable materials. These collaborations are essential for addressing global issues such as climate change and advancing drug discovery.

In the healthcare sector, AI is aiding in the early detection and treatment of diseases like cancer and improving radiological practices.

In the realm of education, AI-powered tools, like **Shiksha Copilot** in India, are helping teachers create lesson plans more efficiently, leading to improved educational outcomes.

Projects such as "Find My Things" demonstrate how AI can cater to personal requirements, emphasizing a people-oriented approach to the advancement of technology. These initiatives, in addition to the Accelerate Foundation Models Research program, showcase the immense potential of AI in addressing societal issues by incorporating knowledge from different fields.

Another good example of successful AI use is the story of Uber. This company has had a major impact on the transportation industry. Uber's use of AI for ride-matching, route optimization, and dynamic pricing is a prime example of how AI can fuel groundbreaking innovation. Travis Kalanick, one of the co-founders of Uber, focused on implementing a strategy centered around artificial intelligence. This strategy has had a profound impact on the way urban transportation functions.

Amazon is a great illustration of how AI plays a prominent role in the platform's economy. The company's implementation of AI for personalized recommendations has revolutionized the online shopping experience, enhancing its user-friendliness and efficiency. Jeff Bezos, the visionary behind Amazon, understood the immense possibilities of AI in revolutionizing different areas of the company, such as supply chain management and customer interaction. The strategic integration of AI has enabled Amazon's growth and achievements.

*"Installing emotional intelligence in AI is like trying to teach a toaster to cry over burnt bread." ~ Dr. Silica Sand, Future Funnies*

## Concerns About AI's Impact on Social Interactions and Relationships

The impact of AI on ethics and society is intricate and diverse. The influence of AI on social interactions and relationships is vast and far-reaching.

There is a lively debate surrounding the influence of AI in education and how it affects human interaction. Experts, like MIT professor Sherry Turkle, have raised concerns about the potential impact of AI implementation on human connections in educational settings. On the other hand, there is a different viewpoint that sees AI as a possible solution to the problems we face in society, rather than something to be feared. Satya Nadella, CEO of Microsoft, believes AI can enhance human creativity and help us solve complex problems.

Supporters of AI in education argue that this technology can address aspects of human well-being rather than isolating individuals. Advocates argue that AI has the power to address feelings of isolation, boost self-assurance, and improve mental well-being and emotional understanding.

A study conducted by the National Library of Medicine explored the effects of the COVID-19 pandemic on college students in the USA. The findings revealed a notable decline in mental well-being among students, with higher rates of depression, anxiety, and stress. These challenges have resulted in increased feelings of loneliness and social isolation. AI has the potential to change how students interact with their peers and educators, making it more engaging and stimulating. Take a look at AI-powered chatbots that offer personalized support and guidance. They can boost your confidence and keep you engaged in your studies. Artificial intelligence can also assist educators in tailoring learning experiences to individual students, fostering a sense of connection between students, teachers, and their peers.

Although AI can enhance student engagement and interest, it cannot replace in-person communication. Students can use AI to create opportunities to interact with teachers and other students and organize live events. For example, you can use AI-powered social networks to connect you with other individuals who share common interests and hobbies, fostering a sense of belonging and reducing feelings of isolation. **(AI and Ethics)**

AI can also help people manage their mental health by providing tailored recommendations and support. Several AI-powered chatbots and robots are designed to help with mental health issues and alleviate loneliness. One such chatbot is **Replika**, which uses machine learning algorithms and natural language processing to create a personalized chatbot that can engage in deep and meaningful conversations with its users.

## Tool or Weapon?

To give a balanced perspective, let's look at recent AI incidents that have sparked concerns.

One of the major ethical concerns with AI is the potential for bias and discrimination. AI systems can perpetuate and amplify existing societal biases, leading to unfair treatment of certain groups of people.

In March 2016, Microsoft released an AI chatbot named **Tay** on Twitter. Tay was designed to learn from interactions with other Twitter users and develop conversational understanding. Within 24 hours, Tay began to post inflammatory and offensive tweets, including racist and sexist comments. Microsoft had to shut down Tay after it started spewing a series of lewd and racist tweets. The incident highlighted the potential for AI to perpetuate and amplify existing societal biases, leading to prejudices and unfair treatment of certain groups of people.

AI is a reflection of humanity, as it is biased in the same way as the data it's trained on. If the data contains biases, as seen in the case where Tay was pulling from all the hate speech that ran on Twitter (X) at the time, the AI system will learn and perpetuate those biases. Developers can develop an awareness of these biases and take steps to mitigate them.

Another ethical concern is the potential for AI to infringe on privacy rights. AI systems can collect and analyze vast amounts of personal data, which can be used to track and monitor individuals without their knowledge or consent. One example of an AI incident that raised privacy concerns is the **Cambridge Analytica scandal**. In 2018, it was revealed that the political consulting firm Cambridge Analytica had harvested the personal data of millions of Facebook users without their consent and used this data to influence the 2016 US presidential election. The incident highlighted the potential for AI to collect and analyze vast amounts of personal data, which can be used to track and monitor individuals without their knowledge or consent.

In 2024, as the U.S. gears up for another election year, the extensive utilization of AI is bound to uncover additional risks associated with this technology. As an example, the technology used to create a **"faceless"** (an avatar or created image used in place of your actual self). YouTube channels can also replicate your voice and image without your consent or awareness. A simple 15-second clip you posted of you speaking can be used to clone your voice. Imagine the power of technology that lets you seamlessly insert yourself into any scene, wear any outfit, and adopt any style with only one click. Or the technology that, using a single photo, will allow you to do entire videos with human-like movements and cloning of your or anyone's voice. These same technologies can also land you in a fabricated scenario that could potentially be incriminating or, at the very least, embarrassing.

The alarming concerns with these technologies were instrumental in sourcing the two notorious strikes in 2023, one by the Writers Guild of America and the other by the Screen Actors Guild (SAG-AFTRA), bringing Hollywood production to a dead stop for most of that year, reminiscent of COVID.

While there is more than just sunshine and happiness in the AI movement, consider that the more you focus on something, the more it expands. Humanity will greatly benefit when more individuals direct their attention toward utilizing AI as a tool, including using AI to protect against weaponized AI.

AI also brings up thought-provoking philosophical questions regarding the importance of human judgment in decision-making processes and the impact of AI on our society. (**Harvard**). In the future, as AI systems continue to evolve, they will possess the capability to make decisions that were once exclusively within the realm of human expertise.

AI has the potential to bring unprecedented benefits to society while simultaneously posing substantial risks. We need to address these ethical concerns and ensure that AI is developed and used responsibly and ethically. This requires collaboration between policymakers, technologists, and other stakeholders to develop ethical guidelines and standards for AI development and deployment. By doing so, we can harness the power of AI to create a better future for all of us, as demonstrated in the UNESCO case studies that we discuss below.

## Historical reaction to technological advancements

The use of calculators and other technological breakthroughs has historically sparked debate. At first, educational institutions were wary of calculators, linking them to cheating and fearing that they would hinder students' comprehension of basic math. Finding a middle ground between enhancing instruction with calculators and preventing students from becoming unduly reliant on them allowed for changes in earlier perceptions. As time went on, it became evident how beneficial calculators were for teaching, particularly in areas like physics and engineering, where they were particularly helpful in helping students understand mathematics more thoroughly and solve difficult issues.

Comparably, navigation has evolved as a result of mapping technology advancements, moving from paper maps to digital tools like GPS and Google Maps, which have caused controversy, while also simplifying travel. Even though digital maps are more practical and offer real-time updates, others argue that they impair map reading and spatial awareness—skills that are essential in situations where technology could break.

Cell phones have revolutionized communication and accessibility, but they come with challenges such as concerns about etiquette, privacy, and reduced face-to-face interaction. Despite these issues, cell phones have enabled relationships to thrive even in remote situations, as seen during COVID lockdowns, though their impact on the quality of relationships and social skills remains a topic of debate.

Initially considered luxury items due to their high cost and limited availability, cell phones became more accessible over time, reflecting a shift similar to OpenAI's democratization of AI technology in 2023. In the case of cell phones, there were many instances where younger people helped older generations bridge the gap, with an unexpected benefit of strengthening family bonds.

Despite the challenges they presented at first, mobile phones have evolved into an integral part of our lives, fulfilling functions that go beyond simple communication. It's important to find a middle ground between embracing the benefits of new technology and tackling any potential negatives. This balanced approach ensures that technology enhances rather than complicates our lives while preserving skills and values.

Historical examples like resistance to calculators, skepticism regarding digital mapping, and resentment towards cell phones reflect humanity's recurring pattern of initial reluctance to accept new technology and eventually embracing the progress. These stages of resistance ultimately lead to integration and acceptance, unlocking the full potential of technology for improving human experiences.

## Rethinking AI Beyond Simply Another Technological Leap

While many specialists view AI as yet another technological advancement that initially faces resistance and ultimately improves lives, many others disagree with this perspective. For example, Kai-Fu Lee discusses the unique and potentially more disruptive nature of artificial intelligence compared to past technologies (**AI Superpowers: China, Silicon Valley, and the New World Order**).

Lee acknowledges that historically, technological progress, like the steam engine and electrification, has led to greater wealth and more jobs, even as some types of work were eliminated, mostly because that technology was specifically designed to eliminate costs by removing jobs away from a group of limited highly skilled workers and simplifying the work to create numerous jobs for low-skilled labor. By contrast, AI is designed to replace much of human work, so this historical argument proves to be unhelpful in predicting future job loss and new jobs creation.

Unlike previous technologies, AI can automate complex cognitive tasks in addition to manual or routine work. This capability presents an exceptional shift in the labor market, potentially making many jobs unnecessary. The workers who remain may find their labor valued less and less, and their tasks will become more focused on managing AI.

A critical point in Lee's book was the prediction that by 2030, the major productivity gains from AI, about $15.7 trillion, would mostly benefit the United States and China. Moreover, within these nations, the wealth would likely be concentrated among a small group of individuals. This forecast suggests a future marked by striking economic inequality, driven by AI's unique characteristics as a general-purpose technology. As Lee's book is over 5 years old, decades in technology time, we now see how third-world countries and smaller countries are moving faster with technology and gaining tremendous speed within their countries. The new rules of the game are how fast you can embrace, integrate, and leverage AI. For example, Japan is clearly winning in robotics, and the EU is leading AI regulation.

In contrast to previous technological revolutions, AI's emergence could redefine the value and role of human labor in the economy. This change raises questions about wealth distribution, job security, and the essence of work in an AI-centric world. Lee's perspective emphasizes the desire for a deeper understanding of AI's distinct impact, preparing us to face the challenges and opportunities ahead.

Although AI has far-reaching ramifications that are greater than those of other disruptive technologies like cell phones, digital mapping, and calculators, its impact is largely dependent on its ethical and responsible use. This may be compared to a hammer, a great tool when used to do work, but a dangerous weapon in the hands of a violent criminal. AI functions in a similar way. When used appropriately, AI is a tool with enormous potential that can spur momentous improvements in many different sectors. However, it can also be used for malicious purposes. As AI continues to affect the future of our world, we must ensure that it is properly regulated and guided by ethical principles and thoughtful considerations, so it remains a force for good.

It becomes evident that the trajectory of AI's impact is difficult to predict. It will take deliberate adaptation, moral governance, and ongoing learning to successfully navigate this future. Embracing AI as a tool for complementing human capabilities rather than a foe can lead to a future where technology and humanity coexist in harmony.

## Summary

We discussed opposing views on AI, showing it as both an ally and a threat. While some industry leaders emphasize AI's revolutionary potential, visionaries like Elon Musk and Stephen Hawking have voiced concerns about the technology's potential to transcend human intelligence and pose existential threats. According to Google CEO Sundar Pichai, AI is profound and has the potential to greatly improve people's lives. We further explored this 'AI paradox,' where many who commend the technology also warn against its unchecked advancement because of potential risks. This duality points to the importance of cautiously advancing AI to guarantee that technology conforms to human values and reduces the impact of undesired consequences, including job displacement or moral quandaries, as demonstrated by the recent layoffs linked to AI integration across multiple large firms.

## Actions to Take:

1. Students: learn to notice the patterns, plan, and take action. One of the most valued business skills, according to Tony Robbins, is being able to spot patterns and know how to react to them once they are identified. The reason behind AI's creation is to replace human tasks. When you look at it in combination with layoffs, you see it as a CLUE. If anxiety and dread are present within you, let them be your motivation to face challenges head-on and embrace the opportunities that this technological revolution will bring. Observe your surroundings as you go through your college and collaborate with AI to devise a plan of action for how you will pivot, and then implement that plan.
2. Working professionals: study, learn, and use AI in your job to increase your value, extend your career longevity or ease the transition if you get caught in a layoff.
3. Everyone: start a side hustle. Use AI to give you suggestions for new ideas and how to carry them out. More details on how to do this are in the last chapter.

Next, we'll cover how AI doesn't fit into college and why some schools want it banned, similar to the way they used to prohibit calculators. We'll also show you how to use AI undetected.

*"The potential benefits of artificial intelligence are huge, so are the dangers."*

*~ Dave Waters, supply chain management*

"AI is like a teenager. It thinks it knows everything, but it still can't do its own laundry." ~ Dr. Maxine Turing, Tech Tidbits Daily

# Chapter 3
## AI in Your Academic World

*"AI is not a substitute for human intelligence; it is a tool to amplify human creativity and ingenuity." ~Pedro Domingo Moreno, a computer scientist*

Approximately "84% of our teachers agree that it is very likely students will require AI skills to succeed in the future workplace," (Oxford University Press), "78% of parents believe that using AI writing tools for schoolwork is a form of cheating," (Turnitin), and "54% of students believe employing AI for academics or studying for tests constitutes cheating or plagiarism," (Best Colleges).

According to a **May 2023 survey**, 90% of high school and college students, totaling 3,000 participants, expressed a strong preference for studying with ChatGPT rather than a human tutor. Additionally, an overwhelming 95% of the students reported an improvement in their grades after utilizing AI for their studies. These findings highlight the engaging and effective nature of ChatGPT as a learning resource.

A majority of colleges have embraced the use of AI, with over 80% either using it or in the process of integrating it. Interestingly, all of the colleges that utilize AI employ it for applicant review and AI detection. Even with the high use, as of May 2024, very few U.S. schools have published their AI usage policies, nor do they provide AI classes for use or AI ethics for their students.

According to the survey that we conducted before writing this book, 44.4% of you are concerned that, due to AI, what you are studying will be irrelevant by the time you graduate. At the same time, 66.7% of you rate your current understanding of Artificial Intelligence as "beginners."

In today's fast-paced world, where technology has become an integral part of our lives, the way we approach education, work, and daily activities is undergoing a rapid transformation. As we step into the future of technological progress and innovation, almost half of students are concerned about the relevance of their degrees when they graduate. They fear being left behind in a world where machines are constantly evolving and learning at a pace that is much faster than humans.

It is important that students develop new skills and educators embrace new ways of teaching. Because AI will likely become the primary way humans access information, professors need to prepare students to use the technology effectively in their lives and careers. One of the valuable skills for students to learn is effective **prompt engineering**, which refers to crafting instructions, directives, and questions that elicit the most useful answers from AI platforms. The more comfortable professors themselves become with using AI, the better they will be at teaching AI skills to students (**aacsb.edu**).

Colleges don't have universal criteria for student AI usage. Some schools restrict AI, while others prohibit it altogether. While some institutions aggressively encourage and assign AI-augmented work, others evaluate AI use for projects individually.

In schools that ban AI, students can learn critical thinking and AI collaboration on their own. However, using AI this way limits personal progress and risks including incorrect assumptions and AI hallucinations. Actively teaching students to collaborate with AI reduces inaccuracies and improves learning.

It's worth looking at an example. Command prompt "Write an essay on how schools could properly arm students for the AI future, write it at a freshman college level, and do it in the tone of a well-spoken revolutionist."

As you can see, from your AI's output, there is zero effort or growth on your part with this prompt. It's also the very reason schools resist teaching and allowing the use of AI. Pure laziness of merely directing your AI robs you of the tremendous power of AI.

---

## PRO TIP

**In order to write originally while using AI, collaborate, provide your thoughts on the subject and ask AI to ask you questions to help draw out your critical thinking. This also enables you to pass AI detection and eliminates risk of " AI hullcinations" and plagiarism.**

---

While you are learning prompting, using GPTs will save time and produce better results. However, if you choose to write your own prompt, here is a command prompt that will produce an original essay that stretches you and has you collaborate with AI. Command prompt "Let's write an essay together on how schools could prepare students on how to survive after graduating in this AI era. It's the school's responsibility to properly prepare me for how to work with and leverage AI. Ask me 5 questions as we write this collaboratively, and refrain from making assumptions. Write it at the freshman college level; do it in the tone of a well-spoken, peaceful revolutionist to convince colleges to allow students the use of AI." This prompt has you provide your opinion to start the AI off in a direction while instructing it to refrain from hallucinating by refraining from making assumptions. The AI will then begin asking you questions, enabling you to go deeper into formulating and clarifying your opinion. If you work with AI in that collaborative manner, schools would more readily agree to its usage, IQ levels would increase, and you would contribute to the survival of humanity.

Currently, Microsoft and Google offer free AI courses. Start there, then do a Google search for "free AI courses from colleges." It's necessary to add "from colleges/universities," or else you will have to wade through the numerous ads and generic courses.

After a few of these courses, you will have concluded what kind of AI you are interested in, from there dive into the prestigious global University of YouTube and search for your AI interest. Congratulations, you just did what your the school has yet to do for you.

Now back to AI in Your Academic World.

Let's consider the influence of AI on your academic and professional futures and how it may challenge you to rethink what it means to learn, work, and thrive in a world where being human is no longer enough to guarantee your place in it.

## Academia

AI has major effects in science and engineering. For example, researchers used AI algorithms to analyze signals from laser interferometers in the LIGO project (the Laser Interferometer Gravitational-Wave Observatory), which resulted in the detection of gravitational waves. This shows that AI can handle difficult scientific data. Boston Dynamics' robots, like Spot, employ AI to negotiate difficult terrain and complete tasks autonomously, demonstrating AI in practical robotics. Biology fans can imagine AI deciphering genetic data to understand intricate living processes.

This demonstrates AI's capacity to process complex scientific data. In engineering, Boston Dynamics uses AI in their robots, like Spot, enabling them to navigate challenging terrain and perform various tasks autonomously, showcasing the integration of AI in practical robotics. Biology enthusiasts can see AI unraveling genetic data, offering insights into complex life processes.

AI is aiding analysis in Humanities and Social Sciences. It helps historians reinterpret old records. For instance, at Stanford University, historians used AI to go through a huge set of diplomatic messages called "Cablegate." This helped them find hidden patterns and connections in history that would be very hard to find by hand. Students studying psychology also find AI very useful for studying how people behave and for getting a better understanding of how the mind works.

AI has many real-world applications for business and economics courses. IBM Watson for market trend predictions, Tableau for data visualization, and Python for data analysis provide meaningful business insights. Financial firms utilize AI algorithms for real-time portfolio management and risk assessment.

Students may predict economic trends, study market behavior, and grasp consumer dynamics with these tools. Integrating AI into academic work helps students learn technology and data-driven decision-making, preparing them for careers in an AI-influenced business sector.

These examples illustrate the profound impact of AI in various academic disciplines, providing students with a glimpse into the future possibilities in their fields.

## Cutting-Edge Research Opportunities

Artificial Intelligence is changing academic research, providing new avenues for exploration and learning across diverse fields. It is becoming an important asset in universities and research centers, aiding experts in enhancing their comprehension of subjects and uncovering fresh insights. The expanding integration of AI in academic research underscores its transformative capacity in reshaping the methods of study and knowledge acquisition.

Let's explore the influence of AI in the medical sector through an examination of IBM Watson for Oncology. This innovative AI system is transforming the healthcare field, particularly in combating cancer, as it collaborates with oncologists to enhance treatment decisions. By efficiently analyzing extensive medical data including research findings, clinical trials, and patient records, Watson generates personalized treatment recommendations for individual patients. Its capacity to equip physicians with up-to-date medical insights is truly remarkable, broadening their considerations. The use of Watson in medicine centers on cutting-edge technology that holds promise for reshaping patient care and offering new opportunities.

AI is changing the field of computer science and engineering, opening up a world of endless possibilities. For example, the advancements made by Google in the field of AI are truly remarkable. DeepMind, a Google subsidiary, deserves recognition for its creation of AlphaGo. This AI system achieved a remarkable victory against a world champion in the complex game of Go. This accomplishment, once considered impossible for computers, demonstrates the capability of AI to effectively tackle intricate problems.

Microsoft has made sizable investments in the field of AI, especially in software development areas. AI has transformed programming tools, making code writing processes more efficient for developers. For instance, Microsoft's Visual Studio Code offers advanced AI features that greatly enhance the coding experience.

Enterprises such as Darktrace utilize AIe to strengthen defenses against digital threats by detecting unusual network behaviors and proactively preventing cyber intrusions.

The incorporation of AI is changing the field of robotics. Boston Dynamics utilizes the power of AI in their robots to tackle a wide range of tasks, from streamlining warehouse operations to carrying out life-saving search and rescue missions. These robots showcase impressive skills in maneuvering through intricate surroundings and carrying out risky tasks on behalf of humans.

These examples highlight the wide range of uses for AI in computer science and engineering. It can make programming tools smarter, improve cybersecurity measures, and drive progress in robotics technology.

AI has become an invaluable partner in environmental science efforts, providing innovative solutions to aid in conservation projects. Wildbook stands out for its use of advanced AI technology to quickly and precisely identify animals from images. This is a valuable tool for researchers who are committed to monitoring endangered species. This tool provides researchers with accurate data that is essential for successful wildlife preservation efforts.

Microsoft's "AI for Earth" initiative uses the power of AI to tackle pressing environmental issues. It focuses on predicting regions at risk of illegal poaching and deforestation, among other challenges. This proactive strategy is incredibly effective in protecting delicate ecosystems, showcasing the immense power of AI-driven approaches in environmental preservation efforts.

AI is also playing a role in the field of marine conservation. Organizations like OceanMind are using artificial intelligence to track fishing operations all over the world and analyze satellite imagery. The prevention of illicit fishing, the promotion of sustainable fishing methods, and the protection of marine life are all aided by this technology.

AI is a valuable asset in climate research. By processing vast amounts of climate data, AI helps scientists understand weather patterns and predict future climate change. This includes forecasting extreme weather events, which is pertinent for preparing and mitigating potential impacts.

Additionally, AI contributes to managing air quality. Projects like IBM's Green Horizon use AI to predict pollution trends and suggest strategies for reducing air pollution. Such applications are essential to improving urban living conditions and promoting public health.

In environmental science, AI is used in everything from protecting wildlife to studying the climate, changing the way we understand and protect the environment. AI is becoming an essential tool that helps us learn more about the world and protect it.

In the fields of psychology and behavioral studies, AI is having a pronounced effect by helping us understand and treat mental health problems better. One great example is Woebot, an AI-powered robot that was made to help people with mental health issues. Woebot helps people with mental health problems right away by giving them tools and interactive conversations. Using machine learning techniques, this chatbot can tell how people are feeling and respond in a way that fits their needs, making mental health help more accessible to everyone.

AI can be used in psychology in many different ways, much more than only chatbots. Researchers are using AI to examine sizable datasets of behavioral trends. This makes it easier to find patterns and possible causes for mental health conditions. This skill to analyze is key for early intervention plans and treatment planning. AI finds insights that were hard to find or didn't exist before by processing data from a variety of sources, such as polls, social media posts, and wearable tech inputs.

Also, new AI tools are being made to help therapists and counselors do their jobs better in therapy and counseling settings. These tools look at how people talk and write during therapy sessions. This helps therapists learn more about their patients' emotional health and growth. This personalized approach helps therapists make interventions that fit each person's needs, which eventually makes therapeutic interventions more effective.

AI is also used in the study of developmental disorders such as autism. By analyzing behavioral and speech patterns, AI systems can assist in early detection and intervention, which is vital for effective treatment.

Likewise, AI can be applied in neuropsychology. AI is used to understand brain function and structure, helping in the diagnosis and treatment of neurological disorders. AI algorithms can analyze brain imaging data to detect abnormalities or changes related to conditions like Alzheimer's disease or stroke.

AI is used in psychology and behavioral sciences, where it creates new ways to study, diagnose, and treat mental health problems. AI is making huge strides in the areas of mental health and behavior analysis by giving them better tools and more detailed analyses of behavioral data.

When AI is added to the art and creative industries, things change in amazing ways, which we talk about in more detail in Chapter 5. Projects like Google's **Magenta** are examples of how AI is used in the arts. The movie Magenta is about AI making art on its own and AI working with human artists to make art in new ways. Magenta pushes the limits of traditional art methods by using machine learning to make original music and visual art.

AI has an effect on art in more than one way besides Magenta. IBM's **Watson** is another example. It was recently used to make a movie clip. Watson looked at hundreds of horror movie trailers to find trends and elements that tend to catch people's attention. It then used this information to edit a new trailer. This experiment showed that AI can understand and copy the emotional effect that movies intend to have.

AI is also making its mark in the world of writing. There are AI programs that can help you write books by coming up with plot twists, character developments, and even whole chapters. Even though these tools are still very new, they open up exciting new ways to write and tell stories.

The field of planning is also being changed by AI. For example, Autodesk's AI-based tools help builders and designers make their designs better in many ways, such as by making them more energy efficient and stronger. These tools allow creators to try out new ideas that weren't possible before by giving them simulations and predictive analysis.

These examples show how AI is becoming more prominent in the creative and art fields. AI is changing what's possible in art, music, design, and film by combining technology with human creativity. It's also changing how we enjoy and make art.

Programs like LIGO, which were already talked about, have helped the fields of physics and astronomy with artificial intelligence. When it comes to studying things that happen in space, AI programs are very useful for looking at very large sets of data that come from finding gravitational waves. Making this happen helps to find things faster.

In these real-life examples, we can see how AI is changing the way academic study is done. You, as students and researchers, are on the verge of a new era in which working together with AI entities, companies, and initiatives will open up the door to unimaginable possibilities and help you learn more in your fields. The combination of human intelligence and AI is bringing about a new era of exploration and achievement, unlike anything we have seen before.

Let's look at AI's transformative role in personalized learning experiences.

AI's role in higher education is evident in platforms like **Coursera** and **edX**. These systems use AI to suggest courses based on your interests and previous learning, creating a custom educational path. This personalized approach helps students navigate through a vast array of academic options, making their educational experience more focused and relevant.

For subjects requiring specific skills, like mathematics or coding, AI tools such as **Wolfram Alpha** and **Codecademy** offer invaluable assistance. They break down complex topics into digestible, manageable steps, making learning less daunting and more intuitive.

In the realm of research and writing, AI tools like **Grammarly** and **Quillbot** are indispensable for students. They offer suggestions to refine writing skills, focusing on grammar, style, and content. These applications help streamline the study process and make learning more efficient.

## PRO TIP

Use AI to create study summaries, quizzes, and flashcards. Have it grade and rank your knowledge and create a custom teaching to boost your weak areas. Know where you stand before the exam.

AI's ability to adapt to individual learning styles is transforming education. Tools like **Khan Academy** use AI to tailor lessons to each student's pace. They ensure that each student progresses at their own pace, thoroughly grasping concepts before moving forward. Adaptive tutoring systems, like **Carnegie Learning**, provide real-time feedback and personalized study paths, making learning more efficient and effective.

These systems are about delivering content and understanding each student's unique learning journey. AI in this context acts less like a teacher and more like a companion that guides and supports students through their academic challenges.

AI's impact extends to language learning as well. Platforms like **Duolingo** use AI to tailor vocabulary and grammar lessons, making the process of learning a new language more engaging and less intimidating. You can even ask ChatGPT or any of the competing AIs to create lesson plans for you to learn any language. It will go back and forth in both text and verbally and you can even ask it to correct you along the way, cost, FREE. Cheat code: have Google translate on the fly IRL, again lazy and you stunt your growth in that area. Works in a pinch.

Ultimately, AI in education can act as a learning companion, adapting to individual styles and needs. In addition to delivering content, it's about understanding and supporting each student's unique academic journey.

## Summary

In this chapter, we examined the growing significance of AI in a variety of professions and highlighted areas in which AI-related skills are becoming increasingly in demand. We saw how AI talents may be utilized in different contexts, in addition to preparing for specific employment requirements.

By acquiring AI skills, you will be preparing yourself for a variety of future professions and ensuring that you are versatile in a global environment that is constantly changing. The development of the prowess to think quickly on your feet and confront challenges head-on is another useful skill to develop. In addition to having a strong understanding of AI, these skills will provide you with a firm foundation that will enable you to adapt and succeed as the world continues to change.

## Actions to Take:

1. Create a LinkedIN account or revamp your existing one using AI.
2. Look at the section of Chapter 10 that covers using AI to find work.
3. Check out additional materials on **DataCamp** for further learning.

4. **Download 500+ Free AI Tools**.

422 million people were victims of or impacted by cybercrime in 2022. In the next chapter, we will give you tips on how to protect yourself.

*"A.I. will force us humans to double down on those talents and skills that only humans possess. The most important thing about A.I. may be that it shows us what it can't do, and so reveals who we are and what we have to offer." ~New York Times*

*Why did the AI student bring a ladder to the art class?*

*To reach the next level of creativity!*

# Chapter 4
## The Importance of Critical Thinking and Creativity

*"Imagination is more important than knowledge. For knowledge is limited, whereas imagination embraces the entire world, stimulating progress and giving birth to evolution." ~ Albert Einstein*

AI scored in the top 1% for original creative thinking, according to a **University of Montana study** published in April 2023. By giving AI systems a common test of divergent thinking, a key component of creativity, they were able to achieve this ground-breaking result. The study poses questions about the changing landscape of employment, education, and art and where the capabilities of AI may complement or surpass human ingenuity.

In this chapter, we look at the pivotal role of critical thinking in navigating the complexities of an AI-dominated world and how it has become a necessary skill as college students get ready for a future where technology will shape everything. We will explore strategies for fostering and enhancing critical thinking skills, emphasizing their importance in an era increasingly defined by automation.

## Critical Thinking and Education

"The paradox of education is precisely this—that as one begins to become conscious, one begins to examine the society in which he is being educated. The purpose of education is to create in a person the ability to look at the world, to make decisions, to say, "This is black or this is white," and to decide whether there is a God or not. To ask questions of the universe and then learn to live with those questions is the way he achieves his own identity." ~ James Baldwin

We would gamble to say that being ordinary is a slow death sentence for many. If true, why do so many fight to "fit in" by choosing to rebuke critical thinking and the work that comes with it? A phrase often heard from your generation is "I don't know." One of Sophia's mentors used to say to her in response to "I don't know," "Well, if you don't know what you think, then who does? I'll ask them."

You can blame the schools for prohibiting or limiting the use of AI. You can blame the government for refraining from instituting AI regulations way back when, or even now, as if that would have done anything more than delay the inevitable. You can also blame the government for not yet developing Universal Basic Income or some other failsafe for the impending massive job restructuring. You can blame businesses for seeking to maximize profits on AI and feverishly building new tools without considering ethics or the need to avoid unintended consequences. You can blame the older generations for messing it up and getting it wrong. OR, you can be the humanifestor that you are, take responsibility, and save yourself. You are an adult now. Some of you will become business leaders, others will run the government and direct our schools. Get moving, get in action, and get results.

## Strategies to Maintain and Enhance Critical Thinking Skills for College Students

Think of your education as an ever-changing fusion of many approaches to learning. Interactive discussions ignite critical arguments and test your viewpoints, while traditional lectures provide the framework with the necessary information. Then, you will get a taste of the theory's real-world applications through hands-on projects. This blend will keep you interested in school and will help you become better at analyzing and evaluating complicated ideas.

Like putting together a puzzle, critical thinking requires you to first examine the information you get with an eye for biases and assumptions. Algorithms can subtly affect the information we get. It matters to recognize these biases. To avoid blindly accepting biased data and content, you need to learn how to critically evaluate information.

Remember that AI systems, despite their advanced nature, have the potential to unintentionally reflect biases. A common source of these biases is the data used to train AI algorithms. AI learns from modeling humans, all of whom have biases.

As an illustration, consider an AI recruitment tool that was trained on past hiring data and is likely to have an unconscious bias toward particular demographics. When it comes to candidate selection, the AI can unwittingly propagate these biases. As a learner, try and delve deeper than the surface of AI-generated content to understand where it came from and to be aware of any potential biases it might have.

Another issue with AI is that it can easily create content that looks real when it's actually made up, whether on purpose or by mistake, through so-called "AI hallucinations". It is therefore useful to develop the habit of continuously questioning the information you come across, whether video, writing, or photos, even if it seems true at first and especially if it validates your own biases and preconceived notions.

It is important to maintain a vibrant sense of curiosity. Dive into the world of AI to understand the technical workings of an algorithm and to explore its broader implications. Why was it created? What problem does it solve, and how might it affect society? Pursuing questions like these transforms your learning experience from a passive intake of facts to an active exploration of knowledge. It's about learning, connecting the dots, and uncovering the bigger picture.

One of the best ways for a college student to prepare for the real world is to participate in activities that require you to solve problems. You can participate in coding competitions or hackathons, frequently hosted by IT companies and student organizations. These events put you in the thick of intense problem-solving scenarios where you devise solutions under time constraints. Another option is to work on group projects that mimic actual business challenges, such as creating a sustainable city plan or coming up with a marketing strategy for a new product. Oftentimes, you could form teams to complete these projects, forcing you to communicate, collaborate, and solve problems.

Another excellent tool for developing your analytical aptitude is case studies. Many courses, particularly those in the social sciences and business, make use of case studies as a means of illustrating complex real-world scenarios. By carefully examining these scenarios, you will learn to identify issues, prioritize solutions, and understand the consequences of various decisions.

Embrace learning that crosses disciplines for a well-rounded intellectual experience. For example, if you are planning to study computer science, taking a psychology class might help you comprehend how people's actions influence the creation of new technologies. If you are studying environmental science, taking a course in economics could give you new insights into the financial implications of environmental legislation. This cross-disciplinary exploration does more than simply broaden your knowledge base; it enhances critical thinking and applies your learning in diverse situations.

Engaging in problem-solving activities and interdisciplinary learning help you to develop a versatile and agile mind. These experiences equip you with the tools to tackle complex problems, encourage innovative thinking, and prepare you for a future where adaptability and creativity are key.

---

## PRO TIP

Practice analyzing and evaluating information critically, skills that AI cannot fully replicate, yet.

1. Ask questions
2. Identify biases and assumptions
3. Evaluate evidence
4. Analyze arguments
5. Consider multiple perspectives

*"Believe nothing, no matter where you read it, or who said it, no matter if I have said it, unless it agrees with your own reason and your own common sense." - Buddha*

# The Vital Role of Critical Thinking in an AI-Driven World

In today's world, where AI and automation are everywhere, to adapt and survive, college students will want to think critically. This skill is key for your studies as well as preparing you for future careers. As routine tasks become automated, the human capacity for adaptive problem-solving, ethical decision-making, and creative innovation becomes more valuable than ever.

While AI is great at automating routine jobs, it still requires human intervention when solving complex problems. Possessing strong critical thinking skills enables you to tackle issues with adaptability and originality. Consider a scenario where AI offers recommendations based on data. In such a case, your level of critical thinking skills will determine if these suggestions take into account real-world complexities and whether they lead to beneficial conclusions. This skill is essential for those working in fields where AI is still in its infancy, such as engineering and healthcare, where humans regularly encounter unexpected challenges.

There is an infinite quantity of knowledge available to us in this digital age. You may make sense of this deluge of information, find reliable sources, and integrate it all with the help of critical thinking. When doing research for an academic paper, use critical thinking skills to identify credible sources and strong arguments. This will guarantee that your work is based on correct and well-examined facts.

Critical thinking also allows you to address the numerous ethical challenges related to technological advancement. For example, taking multiple perspectives into account and examining the potential repercussions of technology-driven decisions need to become a necessary part of creating any new software.

Being able to continuously learn new things and enhance one's skills is essential in a world where Ai and other technology is ever-changing and creating new job opportunities. This mindset is essential for those who work in fields like data analytics or digital marketing, where being up-to-date with the latest methods and technologies is key to success.

Creativity and critical thinking are complementary. There will be more room for original thought as AI takes over more mundane jobs. As an example, AI has the potential to revolutionize graphic design by freeing up designers to concentrate on the more creative parts of their job by automating basic layout duties. Innovative problem solvers with a critical mindset can use AI to uncover previously unanticipated opportunities and insights.

Remember that, although technology is constantly improving, the ability to think critically is unique to humans. If you want to succeed in a world where AI rules, you need to be able to adapt and use these skills to your advantage. After all, technology can only help human creativity and innovation to go so far.

We are entering a new age of artificial intelligence, and one thing that is immediately apparent is that AI is a catalyst for creative inquiry, pushing innovators and artists to push the frontiers of what is considered possible. The examples below show how AI is transforming various artistic domains while underscoring the irreplaceable value of human imagination and ingenuity.

As a fascinating new collaborator, AI is capturing the imagination of artists. One example is **DeepDream**, a Google product that employs neural networks to create surreal artwork from regular photos. In a similar vein, AI algorithms have been employed to generate novel visual art forms. For instance, the **Next Rembrandt** project utilized AI to examine Rembrandt's paintings and generate a completely new work in his style. Thanks to these developments, which combine conventional techniques with AI's computational imagination, creative expression is being redefined.

**Amper Music** and **AIVA** (Artificial Intelligence Virtual Artist) are two examples of AI systems that are revolutionizing the music industry. AIVA can sift through an extensive catalog of classical music to generate new compositions, offering composers new ideas and viewpoints. This system is capable of comprehending and implementing intricate music theories, enabling it to produce works that are inventive and technically solid.

Amper Music does things a little differently. It is a program that helps producers and artists make music quickly and tailor it to their needs. Users can obtain a composition that AI customizes to their unique requirements by inputting style, mood, and length. When working with limited time and money, this technique shines for scoring commercials, sports, or movies.

Poetry, short stories, and even research papers are all areas where AI is leaving its imprint on the literary world. ChatGPT and **Gemini** are only two examples of apps that show how AI can write stories and poetry, sometimes even in the style of well-known writers. With the use of prompts, ChatGPT can produce text, such as stories or poetry. If you look for help with writing academic papers or other organized pieces, Gemini is the way to go. Writing out parts of a paper, structuring ideas, or even coming up with references are among the tasks where it can help. Because it simplifies writing and generates new ideas, this can be a blessing for students working on difficult research topics.

AI in music and literature is about enhancing human creativity. No more "writer's block." These tools provide artists and students with a starting point from which they can expand, experiment, and express their unique ideas. By utilizing AI, musicians can explore new sounds and compositions, while writers can enrich their narratives. As a collaborator in both domains, AI contributes a unique combination of technical expertise and imaginative capacity. The potential of AI in the arts is only scratching the surface of its capabilities. This bodes well for the future of artistic expression.

AI is bringing new ideas to the field of fashion design. Algorithms that research material qualities, consumer preferences, and trends aid in the development of innovative styles. These AI systems can enhance the designer's creative process by suggesting new patterns, color combinations, and even future fashion trends. As an example, a creative might use AI to research what kinds of materials and styles are now trending, then mix that information with their own artistic vision to make something totally original.

Robotic intelligence is revolutionizing the film industry. AI script writing systems can take genre trends and audience tastes into account when making storyline or character arc suggestions. Instead of replacing them, this gives screenwriters a plethora of possibilities to think about. AI audience prediction tools also provide filmmakers with a sense of how their work could be received, so they can tweak it to meet viewer expectations while still being true to their artistic vision.

Deep learning systems, such as OpenAI's **DALL-E**, are pushing the envelope of what is possible in the visual arts. To help artists conceptualize intangible ideas or seemingly impossible situations, DALL-E can produce visuals based on written descriptions, known as "Text to Image." Artists may now realize their wildest dreams with the help of this program, which lets them produce images that were previously unimaginable when rendered manually.

Technologies for synthetic voices, notably those created by **Eleven Labs**, are causing a sea of change in industries such as digital narration and audiobook production. Creators have a versatile tool in Eleven Labs' capacity to generate realistic speech in multiple languages and voices. This tool offers a spectrum of vocal emotions that add depth and nuance to screenplays, bringing them to life.

With **Descript**, editors can devote more time to storytelling and less to the technical aspects of audio and video editing. Similarly, filmmakers and video artists have access to cutting-edge tools for video creation and editing with **Runway** and the newly launched **Sora**, which makes complex visual effects more approachable and controllable.

It is becoming apparent that AI is changing the way we approach art, design, and storytelling as it spreads throughout the creative professions. Offering new avenues for exploration and invention, AI provides a fresh perspective through which to examine and comprehend creative processes. Still, at its core, all creative pursuits revolve around the singularity of the human viewpoint, the nuance of our imagination, and our capacity to forge emotional connections with others. While AI can impress us with "micro-creativity", coming up with original ideas, the realm of "macro-creativity" still belongs to humans.

## Balancing AI Insights with Intuitive Decision-Making

Integrating data driven by AI with your your human judgment is like having a superpower in our modern day.

Consider this basic example: an AI-powered streaming service that learns your watching habits to suggest movies. While the algorithm can recommend movies based on your viewing history, nobody can predict how you are feeling on any particular evening. Therefore, it's your mood and gut instinct that will guide you to make a choice based on your present emotional state rather than the habitual pattern.

Similarly, AI systems in the financial sector can examine past data to predict future market movements. Regardless, subtle changes in investor sentiment or unforeseen economic movements have eluded their predictions so far. In such cases, your intuition and knowledge of up-to-date international news, economic data, and market psychology are pivotal to supplementing predictions made by AI.

Your intuition can act as an internal compass in situations where it's necessary to interpret qualitative data derived from quantitative data provided by AI. The knack to "read between the lines" of a marketing presentation or identify hidden tensions in a group setting is at the heart of this concept. An intuitive understanding of human emotions and reactions might offer deeper insights into the meaning of data provided by AI regarding client preferences or team performance measures.

Academic research greatly benefits from AI's data processing capabilities, especially when it comes to handling large datasets like climate change statistics. Human intuition and analysis are still necessary for making sense of these results and their practical implications. It all revolves around bridging the gap between the qualitative context that human intuition can provide and quantitative analysis by AI.

Making good decisions in most fields depends on balancing human intuition and technological capabilities. It's like being an orchestra conductor, with AI supplying the data (the score) and your intuition (the baton) directing the tempo and dynamics to produce a beautiful symphony of choices. You can make decisions that are based on statistics and also sensitive to people's experiences if you adopt this dual approach.

Creativity and critical thinking have become essential qualities for navigating a technology-dominated society. Students can innovate in their disciplines, adapt to the quick changes brought about by AI, and tackle issues with a fresh perspective by fostering these abilities.

## Summary

We spoke about the pressing role of creativity and critical thinking in an era where AI is becoming increasingly dominant, the importance of imagination over knowledge, noted by Einstein. However, we also looked at the study that casts doubt on the idea that creativity is exclusively human by showing that AI might demonstrate high levels of original creative thinking. This discovery starts a larger conversation about the effects of AI on creative industries. We shared our belief that critical thinking will be a necessary skill in the AI-driven society. To prepare for this future, we encourage you to incorporate critical thinking into your education.

## Actions to Take:

1. When using AI, collaborate with it rather than directing AI by including your thoughts in your prompt. If you are unclear about the subject, request that the AI ask you a few questions to help you articulate your thoughts.

2. Engage in interdisciplinary studies to broaden perspectives, combining subjects like science and art or technology and humanities. For example, a course like "Digital Humanities" at Stanford University blends tech with humanities to analyze historical texts with computational tools, offering new insights into old works.

3. Participate in problem-solving and creative thinking workshops, e.g., those offered by MIT's "Solve" initiative.

4. Practice reflective thinking to question assumptions and biases. You can consider keeping a journal where you write down thoughts on what you've learned, question your initial assumptions, and consider other viewpoints. Harvard's "Project Zero" offers tools and routines for developing reflective thinking. Question everything, especially what you know to be true.

5. Participating in hackathons or innovation challenges, like those hosted by TechCrunch, can be a great way to learn from a potential failure in a safe environment. These events encourage trying out new ideas and learning from unsuccessful attempts.
6. Ask Ai to help increase your critical thinking skills and to come up with games, puzzles, etc. to help you

In the next chapter, we're going to discuss AI replacing 300 million jobs and why we will love that.

> *"It is astonishing what an effort it seems to be for many people to put their brains definitely and systematically to work." ~ Benjamin Franklin*

"You don't have to worry about AI stealing your job—unless you're a calculator."
~ Bob Binary, Technically Funny

# Chapter 5
## Career Preparation for an AI Future

*"I think the future of global competition is, unambiguously, about creative talent, and I'm far from the only person who sees this as the main competition point going forward. Everyone will have access to amazing AI. Your vendor on that will not be a huge differentiator. Your creative talent though – that will be who you are." ~ Vivienne Ming, executive chair and co-founder, Socos Labs*

The World Economic Forum's Future of Jobs Report 2023 projected that 23% of global jobs will change in five years. This estimate of 45 economies and 673 million workers anticipates 69 million new jobs and 83 million losses due to macrotrends and technological improvements, including AI. This statistic underscores the urgency for workforce adaptability and skill development in response to the rapid changes brought about by AI and technological innovation.

In Chapter 3, we showed you how to get around the lack of formal education in AI and where to go to get free education. The next step is to intern. Do this way before you graduate. Find a company or a mentor that has integrated AI and offer your time in exchange for real-world applications. Or, as Sophia would recommend, you may choose to just skip to the end of this chapter and learn how to make money for yourself using AI. As you are interning or working for yourself using AI, you might even reconsider your reason for attending college.

In this chapter, we will go into strategic career preparation for an AI-driven future and look at the necessary skills and insights that you will need to develop to thrive in the evolving job landscape.

## Current Trends in the Workplace

Before making any career recommendations, let's take a look at how the new workplace developments in an AI-driven environment are affecting career perspectives and job engagement. These findings are based on a basic observation: AI and technology are changing job responsibilities and job loyalty.

A new trend of "quiet quitting" has recently developed in our workplace. In our new world dominated by technology, the concept of loyalty is shifting. People are leaving jobs that don't match their skills, professional goals, or their calling. As AI redefines positions, professionals are rethinking their career goals and work engagement.

Another modern workplace trend is social impact. Professionals, especially younger ones, are increasingly seeking careers that reflect their convictions and allow them to make a positive impact on our society. Companies that share these values are more likely to attract and retain talent because more people value meaning over career metrics.

Job security changes too. In previous generations, job stability meant working for one employer for many years or even a lifetime. Security now depends on agility, adaptability, and skill development. It comes from being able to adapt to new technology, negotiate a rapidly changing employment market, and keep learning new skills.

These patterns show that labor is no longer about money. Rather, it's about finding a balance between personal ideals, job goals, and AI-driven economic demands.

## Job Market Transformations in the AI Era

Two other perspectives provide invaluable insights into the transformation of our workplace and daily lives: one highlights the potential displacement of routine jobs by AI, while the other raises concerns about the broader societal implications related to job market changes, including the emergence of a "useless class" due to automation.

In *21 Lessons for the 21st Century*, Yuval Noah Harari explores the profound impact of AI and technology on the job market. He thinks that numerous conventional jobs, particularly those that involve repetitive tasks such as manufacturing or basic customer service, are in danger of being replaced by automation. Harari highlights the remarkable progress of AI in taking over tasks that were traditionally performed by humans. With the advancement of AI, there is a possibility of it replacing human workers in many fields.

Harari also examines the broader implications of these factors on society and the economy. He expresses concern about the potential impact of AI on employment and the challenges individuals may face in adjusting to new job opportunities. This could potentially result in the rise of a group of individuals who are unable to find employment, often referred to as the "useless class." Harari believes that there is a need for a shift in our approach to education and training, placing greater emphasis on the development of adaptability and lifelong learning skills. According to him, this change is necessary in order to navigate the future job market and technological advancements.

*…erpowers: China, Silicon Valley, and the New…*
…) discusses the profound influence of AI on …l risks of job displacement while also …emain secure in this changing …t jobs involving repetitive, …ed by AI.

…that are expected to remain …rtain types of jobs, which require human creativity, strategic thinking, or complex emotional interactions, are less susceptible to automation. There are a variety of roles available in fields such as research, the arts, strategic business, or mental health care.

It's dumbfounding to see how rapidly AI has surpassed Kai-Fu Lee's assessment. AI currently reigns supreme in research and even in the arts, with programs like Midjourney and Dall·E. CoPilot and ChatGPT 4 have gained attention on social media for their ability to help create and grow companies. Meanwhile, the field of mental health AI is making progress with chatbots that demonstrate empathy, identify needs, provide diagnoses, and offer helpful resources. Available around the clock.

In their article "Is AI the Future of Mental Healthcare?" published by *The National Library of Medicine* in May 2023, Francesca Minerva and Alberto Giubilini assert that "AI in psychiatry has yielded positive results for patients, making the dehumanization of psychiatry a cost worth paying—especially if it reduces costs and improves access; in addition, the very nature of psychiatry as essentially grounded in the human connection between the therapist and the patient is now being called into question."

As we write this book, it is important to acknowledge the limitations of AI in job performance. It faces challenges when operating in unfamiliar, unpredictable, or disorganized environments. Due to these existing constraints, certain occupations will continue to rely on human involvement for a longer period of time compared to others. We delve deeper into this topic later on in this chapter.

You can prepare for these new opportunities by studying fields that are likely to be in high demand in the future, such as machine learning, natural language processing, and artificial intelligence engineering. You also to work on developing skills that are difficult to automate, such as empathy, and critical thinking. By doing so, you will position you success in a rapidly changing job market.

# Navigating Job Market Transformations in the AI Era

As we noted above, in the new AI-dominated world, there are certain jobs that are expected to remain secure at least for the immediate future. These jobs rely on human creativity and emotional intelligence, which are qualities that machines have yet to replicate. Jobs in areas such as research, the arts, strategic business planning, and mental health care have a lower risk of being automated. AI struggles in tasks that require precise hand-eye coordination, intricate strategic planning, and the ability to navigate unpredictable or unstructured situations. In addition, the lack of empathy and emotional connection in AI restricts its effectiveness in professions that heavily depend on comprehending and engaging with human emotions. As a result, professions that heavily rely on these unique human abilities are predicted to remain important and sought after.

For example, AI will lead to new jobs in areas like data science, machine learning, and AI engineering. As AI grows, there will be much demand for people to help direct and develop it. This field is always changing, so workers need to keep up with new technology, something that people are good at. These jobs will require specific skills and knowledge that you can learn in school.

AI will also create new opportunities in fields such as healthcare, education, and customer service. For example, AI can help doctors diagnose diseases more accurately and efficiently, which will improve patient outcomes. Among other things, AI systems can analyze scans and patient histories to assist in the early detection of diseases like Alzheimer's or heart conditions, often with greater accuracy than human practitioners. In healthcare, while AI assists in analytical and administrative tasks, the essence of care remains a human endeavor. Compassion, support, and encouragement, the heart of healthcare, are provided by skilled human professionals.

In the fields of psychiatry, social work, and counseling, the essential qualities of empathy, emotional depth, and trust-building will continue to play an essential role. As we discussed above, AI can help us comprehend and connect with the intricate emotional states of people. A psychiatrist needs to diagnose and treat mental health conditions, and also establish a personal connection with patients, empathizing with their individual experiences and emotions. In the same way, social workers offer a level of support that goes beyond just providing procedural assistance. They delve into personal matters, serving as a connection between people and the resources they require from the community. This demands a level of empathy and comprehension that AI has yet to replicate.

Counseling relies on creating a secure and empathetic environment for individuals to delve into their emotions and difficulties. The effectiveness of a counselor lies in their skill to establish a genuine connection with clients, providing them with empathy and tailored guidance that cannot be replicated by AI. The counselor's role extends beyond giving advice; it involves comprehending the intricate emotional context of each person's unique circumstances.

The concepts of sensitivity and dexterity are extremely important in the field of therapeutic activities such as massage therapy and physical therapy. A good example of this would be the close collaboration that physical therapists have with their patients in order to assist them in recovering from injuries or effectively managing long-term diseases. Their profession requires them to have a comprehensive understanding of human anatomy and physiology, as well as a caring and hands-on approach, in order to aid patients in regaining mobility and easing suffering associated with their condition. Because the requirements of each individual can vary substantially, it is necessary to take a specialized and adaptable strategy, which may provide difficulties for an artificial intelligence system to fulfill.

In the same way, the ability of the therapist to provide accurate pressure and movement in order to reduce tension and pain is essential to the success of massage treatment. Skill, a natural understanding of the human body, and a heightened sensitivity to touch are all necessary components. When it comes to successful therapy, the capacity to intuitively understand and connect with other people is absolutely necessary. Unfortunately, artificial intelligence in its current form is not capable of replicating this essential component.

The creative process of humans continues to be the driving force behind fields such as fiction writing. For the time being, artificial intelligence does not possess the capability to duplicate the imagination, storytelling skills, and emotional depth that are the hallmarks of great writing. Even when students are in the tenth grade, education will still have its human element. When students are at this point in their educational journey, it is essential to make learning interesting and engaging for them. Teachers have a significant influence on pupils because they provide them with a source of motivation, direction, and mentoring that is incomparable to what artificial intelligence can do. By enabling educators to design individualized educational experiences for their students, artificial intelligence has the potential to completely transform the educational system.

Individuals' natural curiosity and limitless imagination are the driving forces behind the advancement of science. Artificial intelligence (AI) is a useful tool in this area, and it absolutely shines when combined with these fundamental aspects of human nature. For the purpose of posing meaningful questions and analyzing the results, scientific research is dependent on the skill and interest of the scientists doing the investigation. The application of artificial intelligence is of critical importance in the process of improving data management and analysis, which ultimately results in the research process being streamlined for maximum efficiency.

On the other hand, our comprehension and interpretation are the sources of the most significant discoveries and the most valuable insights. Take, for instance, the field of genetic research. Artificial intelligence is capable of performing rapid analysis of genetic material; nevertheless, it is still dependent on the knowledge and experience of scientists in order to comprehend the implications of this data on human biology and health.

When it comes to the practice of criminal defense law, the ability to possess the requisite skills is genuinely unique. These skills require the capacity to plan strategically, communicate persuasively, and develop trust, which is an extremely valuable skill. In order to be successful in the courtroom, it is essential for a lawyer to have a profound comprehension of the reactions of the jury and the ability to retort to those reactions in the moment. This is an extremely difficult undertaking that is beyond the capability of artificial intelligence. Due to the fact that the practice of legal defense is primarily dependent on the intuition and emotional intelligence of persons, it is a profession that continues to be dominated by human beings.

Human skills are essential in the field of management. Being a successful leader requires the ability to inspire and motivate a team, foster a positive and supportive work atmosphere, and establish authentic connections with employees. These elements of management extend beyond basic task delegation or troubleshooting. They require a thorough comprehension of the emotional and professional needs of individuals within a team. Take, for example, how a manager can identify and cultivate an employee's hidden talents, resulting in improved team spirit and productivity that artificial intelligence simply has yet to match. Effective management relies on the fundamental qualities that are inherent to human beings, highlighting the significance of human engagement in leadership positions, even in the face of technological progress.

When it comes to providing great support to customers, it is essential to take a personal approach during the customer service process. On the other hand, the incorporation of AI technology has the potential to significantly improve the capacity of businesses to deliver customer help that is both superior and more effective. Customer service professionals are particularly skilled at addressing difficult and sensitive issues, whereas technology is responsible for handling regular inquiries. The client experience is greatly improved as a result of their skilled and proficient management of complex interactions, which makes the experience more responsive and genuinely helpful.

Journalism is fueled by the perspectives and storytelling talents of its practitioners. Journalists go beyond the surface of stories, utilizing their expertise to analyze information and craft engaging narratives. The inclusion of a human perspective enhances the quality of news, offering a greater level of depth and context that has yet to be achieved through technology alone.

In the realm of architecture, the ingenuity and vision of architects are of utmost importance. They create structures that are visually striking and environmentally conscious by combining aesthetic vision with sustainable design principles. Technology may provide precision, and it is the architects' creative thinking and design expertise that truly breathe life into buildings, transforming them into functional and inspiring spaces. The importance of personal expertise and creativity is evident in various fields such as customer service, journalism, and architecture. This highlights the invaluable role these qualities play in today's workplace.

These professions highlight the importance of human qualities such as creativity, empathy, and adaptability, which continue to be valuable in a world influenced by AI. (**Medium**)

# Identifying AI-Related Careers

The advancement of AI is changing existing job positions and also creating brand-new careers. As we look ahead to the future, we can see a wide variety of exciting careers that have been shaped by the rise of AI. These careers focus on the seamless integration, responsible oversight, and ethical use of AI technologies, and they are projected to see enumerable growth and demand. Let's delve into a few AI-focused job roles that are gaining importance across different sectors.

*AI Ethicist and Bias Analyst:* Examine the algorithms used in AI to identify and eliminate biases, and ensure that these systems contribute to social equality. This is a growing demand in police, protection, and security lines of work.

*AI System Trainers:* Work with AI models to handle data correctly and ensure that people who lack specialized knowledge can understand the complicated choices that AI makes. Their knowledge is particularly relevant in fields like healthcare and banking, where AI-driven decisions can have big effects.

*Specialists in Human-Machine Teaming:* Make it easy for AI technologies to work with human teams, resulting in increased output and encouraging new ideas. For example, in manufacturing, they might come up with AI systems that help humans find flaws in products, improving quality control and safety.

*Data Privacy Managers:* In charge of making sure that businesses that use AI and deal with large datasets follow safety protocols, ethical standards, and legal requirements. They protect sensitive information from being misused by using strict access protocols and advanced encryption methods, which are especially relevant in financial institutions.

*Robotic Engineers and Technicians:* As society adopts new technologies, the demand for robotics soars. These specialists find it easier to work in a wide range of fields. For example, people work on making robotic arms specifically designed for assembly lines. In the healthcare business, robotic assistants help with surgeries and patient care. Space exploration also has a lot of exciting possibilities, like making rovers and other robot tools that can be used on journeys beyond Earth.

*AI Healthcare Technicians:* Work with advanced AI equipment that looks at medical images to identify, diagnose, and treat diseases and illnesses.

*AI Integration Specialists:* Add AI to existing systems to automate and increase performance and productivity. For example, in the car industry, they might focus on adding AI to supply chain management systems to make resource timing predictions more accurate.

*Customer Experience Design:* Use AI to change how people connect with businesses and how people shop and get services. Many companies look to hire people who know how to use AI to create outstanding shopping experiences for customers. As part of their job, they make AI-powered solutions like service bots that help customers in the best way possible and targeted product suggestions.

*Policymakers and Advisors on AI:* Because artificial intelligence is developing so quickly, it's dire that we quickly formulate rules and laws ensuring that AI grows in a way that best suits humanity. These specialists play a key role in making laws and rules that encourage AI's moral and responsible use by helping governments and foreign groups figure out how to navigate this complex area.

*Sustainable AI Specialists:* Make AI systems more energy-efficient and long-lasting. They might be working on making data centers run more efficiently so they use less energy or creating AI algorithms to help with environmental tracking and conservation efforts.

***Digital Marketing Analysts:*** Use AI to create targeted marketing strategies that get customers' attention. To be successful in this field, it is necessary to be good at analyzing data, understanding how customer behavior changes over time, and using relevant AI tools. To see more on this, look at HubSpot and Adobe, the first companies to use AI in marketing.

***Supply Chain Optimization Specialists:*** Use AI to improve supply chain operations. To be successful in this job, learn how to use AI technologies and also develop data analysis and logistic management skills. Fortune 500 companies like IBM and SAP are among the first to use AI in their supply chain management.

These job paths are exciting for people who want to work in an AI-driven business setting because they combine technology, strategic thinking, and creative thinking. AI's potential to change many different industries opens up a huge number of exciting and rewarding job possibilities that promise growth and advancement.

## AI Development Careers

Getting involved in AI development can help cultivate a sense of ownership and responsibility. Students have the opportunity to go beyond being passive technology consumers and instead become active contributors to its evolution. Here, we explore various ways for students to shift from being passive consumers to active contributors in the exciting world of AI.

Python programming language is highly regarded as the preferred language for AI development. It is known for its user-friendly syntax, extensive collection of AI frameworks and libraries, and impressive data visualization capabilities. With its versatility and ease of understanding, it is perfect for individuals at any level of expertise. There are a wide range of free online courses and tutorials available, offering a practical way for students to improve their coding skills in AI development.

## Qualities Necessary for AI-Resilient Careers

Students in higher education are advised to take a more proactive approach to their education and broaden their horizons in terms of learning to better prepare themselves for future positions of employment. While it is essential to investigate topics such as machine learning algorithms, data analysis, and applications of artificial intelligence in domains such as healthcare and finance, it is equally essential to investigate the ethical implications of human-computer interaction.

Exploring the many facets of artificial intelligence, including its technological complexities and ethical implications, will provide a broad perspective on the enormous impact that AI has on society. Have a look at this amazing initiative that mixes artificial intelligence with concerns about the environment. This presents a wonderful opportunity for you to enhance your technical skills while simultaneously gaining a more in-depth comprehension of the ethical and ecological effects that artificial intelligence can have. To successfully navigate the complicated world of artificial intelligence technologies, it is essential to have a comprehensive understanding of the technology.

To get the best possible results, investigate the symbiotic relationship that exists between humans and AI, as well as the ways in which both can mutually benefit from and enhance the capabilities of the other. There are three primary roles that experts emphasize the importance of humans playing: training robots, explaining the implications of their actions, and ensuring responsible use to prevent harm to humans. Artificial intelligence has the potential to improve our cognitive competency, interact with both consumers and employees, and even expand our physical capabilities by possessing skills that are similar to those of humans.

## Team Dynamics in the AI Era

There are many useful tools on these online learning platforms, and the courses cover a wide range of topics, from the basics of machine learning to the moral issues surrounding AI.

Take **Slack**, for instance. This messaging app has gained popularity in both professional and academic environments. Imagine it as your online gathering space where you can have lively discussions, exchange thoughts, and generate creative ideas. Slack is designed to help you easily manage and stay on top of various conversations by organizing them into channels. Slack revolutionizes team communication by enhancing efficiency and streamlining processes. It enables seamless file sharing, prompt feedback, and effective collaboration, keeping everyone informed and connected.

Virtual collaboration spaces are a compelling application of AI. Platforms such as **Spatial** offer a stimulating experience for students to collaborate, regardless of geographical separation. They provide an interactive virtual setting where team members can work together, using avatars to represent themselves, and easily share 3D content. This is particularly beneficial for projects that demand extensive amounts of visual collaboration, such as in architecture or design.

## Embracing Continuous Learning in a Tech-Driven World

Today's rapidly evolving world of AI and technology is similar to maneuvering a vessel through constantly shifting waters. To stay above water, be flexible, and creative, and develop a mindset where you are eager to acquire knowledge and continuously learn.

Once again, we suggest exploring online platforms such as Coursera and edX, which we covered in Chapter 3, as your digital academies. These platforms provide a wealth of knowledge, offering courses that span a wide range of topics, from the basics of machine learning to the ethical aspects of AI. They offer a user-friendly approach to exploring different tech topics, whether you're new to the subject or seeking to enhance your current understanding. These courses offer a wealth of knowledge and opportunities to explore interesting aspects of AI and technology that may be new to you. They provide a wide range of opportunities, allowing you to explore and learn in a way that suits you best.

Getting certified by big names in the field, like Google Cloud or Microsoft will help you to acquire the necessary tools to succeed in the tech world. Getting each license is a big step forward in your career because it gives you useful skills that make you more marketable. People in your field will give these licenses a lot of weight, which shows how knowledgeable and dedicated you are. Getting certificates in areas like cybersecurity, cloud computing, or data analysis shows that you are dedicated to staying up-to-date in a field that is constantly changing and helps you learn more.

Keep up with the latest advancements in technology by using AI chatbots to give you curated daily highlights on the areas of importance to you, saving you hours. Direct it to pull from blogs and newsletters from tech influencers and visionaries, identify the latest trends, advancements, and overall trajectory of the tech industry. Create your bot to provide valuable insights that can broaden your knowledge of technology and its influence, helping you stay ahead of the curve. Direct your bot to follow influencers on platforms like LinkedIn or Twitter. With an AI bot, you can stay up-to-date with the latest tech developments and engage in real-time discussions in a mere 3-5 minutes daily.

As AI and technology change quickly, learning new things is a bit like steering a boat through water that is constantly moving. Be flexible, creative, and hungry for knowledge to stay afloat and thrive.

As your virtual academies, we suggest that you look into online learning sites like Coursera and edX, which we talked about in Chapter 3. These platforms are like treasure troves of information. They have classes on a wide range of topics, from basic machine learning ideas to the moral aspects of AI.

---

## PRO TIP

Follow industry leaders on social media and engage with their posts to build your network and learn from the forefront of AI development

---

## Networking and Building Your Tech Community

Exploring the vast and interconnected realm of the tech world opens up new opportunities and pathways through strong connections.

LinkedIn offers much more than just a platform for showcasing your resume. It's a platform that allows you to connect with professionals, showcase your work, and gain insights from industry leaders. Every interaction on LinkedIn, whether it's leaving a comment, sharing a post, or making a new connection, contributes to the growth of your professional network. By actively participating in discussions, joining LinkedIn groups relevant to your field, and keeping up with tech companies and influencers, you can drastically broaden your knowledge of the industry.

Attending events such as TechCrunch Disrupt or local tech meetups offers an enriching experience beyond simply sitting through presentations. These events provide a chance to connect with professionals, exchange thoughts, and potentially discover future prospects for employment or collaboration.

Attending conferences, regardless of their size, can provide you with valuable connections and knowledge that can greatly benefit your career.

## Practical Experience: The Real-World Tech Playground

The realm of technology extends far beyond mere academic understanding - it hinges on actively engaging in practical endeavors. There are numerous internships and cooperative programs that offer a chance to learn more about the tech sector. Participation in these initiatives allows you to observe firsthand how technology addresses challenges in everyday scenarios. These opportunities enable the application of theoretical knowledge to authentic professional settings, whether through a summer internship at a tech firm or a co-op role within a global corporation.

Universities provide numerous resources for students who are interested in technology. In addition, schools often organize workshops, seminars, and networking events with alumni, which offer valuable knowledge and opportunities to make connections. These resources are essential for your journey, providing valuable support and connecting you with a wide network of professionals and peers in the tech industry.

To be successful in the technology field, build a solid network and gain practical experience. A striking impact on your career in this dynamic sector can be made by exploring a variety of routes, such as internships and LinkedIn, as well as by taking advantage of chances offered by universities.

## Staying Agile and Adaptable in AI

To achieve success in the fast-paced field of artificial intelligence, make sure to be flexible and agile.

In Chapter 1, platforms like **Kaggle** are highlighted as being much more than places to find datasets and compete with others. They offer a range of interesting and engaging opportunities. They provide a dynamic environment where you can enhance your AI skills, connect with a community of like-minded individuals and professionals, and stay informed about the latest developments and opportunities in the field of AI. Joining Kaggle competitions or exploring their datasets offers a chance to apply your knowledge in practical situations, gaining valuable experience and receiving feedback.

By actively participating in AI forums or becoming a member of university AI clubs, you can fully immerse yourself in the realm of AI. These communities offer a platform for you to express your ideas, gain insights from others' experiences, and stay informed about the latest developments in AI. Getting involved in different communities, like online forums or university clubs, is a great way to keep learning and improving your skills.

There are countless opportunities to become involved in the world of AI and technology. Exploring them allows you to find your niche. Discovering the vast potential of AI in addressing environmental challenges or learning about the complex realm of AI in finance can be both surprising and fulfilling. By finding an alignment between your passions and the AI opportunities, you can create a happy space that could make a remarkable impact.

Establishing a presence on platforms like **GitHub** or sharing your insights on a personal tech blog goes beyond mere visibility; it allows you to shape your distinct identity in the tech world. Sharing your projects or expressing your ideas online can be a valuable way to cultivate your own unique perspective and deepen your understanding of the subject.

Ultimately, learning about technology is a continuous journey that demands continuous education, forging connections, and acquiring hands-on experience. This chapter is designed to assist you in exploring the realm of technology and becoming proficient in your own unique way. It will give you the essential guidance to stay current and leave a lasting impression in this constantly changing field.

## Cultivating AI-Ready Skills for Professional Growth

If you want to be successful in a professional setting that makes use of artificial intelligence, there are a few talents that are important for you to have. *Data literacy*, which refers to the capacity to interpret and work with vast volumes of data, is one factor that jumps out as noteworthy. To illustrate, if you can evaluate customer data, you will be able to aid businesses in acquiring insights into the preferences of their clients, a well-sought-after skill set.

It is also helpful to comprehend the data that AI provides and to make insightful inquiries about it. For example, when using AI in financial operations, examining and interpreting the predictions made by the AI can prevent detrimental errors.

In addition, you will be required to have excellent communication skills, especially when it comes to explaining complex concepts related to AI to people who lack technical competence. Possessing the capacity to convey one's thoughts in a clear and concise manner is very helpful.

By effectively demonstrating how AI enhances customer service in a way that is easy to understand, you can greatly boost the acceptance and use of this technology. By learning these skills, students can position themselves for success in a professional world where AI plays an ever-growing role.

## Path to Success with AI in Various Fields

The rapid progress of AI technology offers many opportunities for making money. AI has the potential to be a valuable source of income, offering opportunities ranging from content creation to advanced coding. In this chapter, we will delve into different methods of utilizing AI to generate income.

AI has transformed the way content is created. It can assist you with writing a diverse array of content, such as blog posts, social media updates, long-form articles, and reports. Incorporating AI into this field opens up exciting possibilities for digital marketers. We recommend that you continue to develop your skills and leverage AI for your success.

AI art generators have ignited a renewed fascination within the art world. These tools offer a unique and innovative approach to creating artwork, allowing for the production of captivating pieces in a matter of minutes. Uncover the secrets of prompt engineering and establish a one-of-a-kind specialty to effectively promote AI-generated artwork on well-known online platforms like Etsy or through user-friendly print-on-demand services. Excelling in this field requires creativity and a commitment to upholding intellectual property rights.

AI code assistants like Copilot can greatly speed up the development of tech products, making them more accessible and efficient. They are capable of automating repetitive tasks, providing code suggestions, and aiding in the process of debugging. This streamlines the process of bringing products to market and enhances the effectiveness of product development. Utilizing these tools can transform a technological concept into a profitable product, as demonstrated by the success of apps like **AudioPen**.

AI tools can streamline various tasks in digital marketing, ranging from creating campaigns to optimizing strategies. This automation enhances the efficiency of digital marketing, allowing marketers to achieve superior outcomes in a shorter period of time. By becoming proficient in AI marketing tools, you can provide businesses with advanced services, catering to the increasing need for digital marketing skills.

AI sales tools bring efficiencies to lead management, sales tracking, and performance analysis. They offer valuable insights that can enhance sales strategies. With a solid understanding of these tools, you'll be able to provide valuable consulting services to businesses. By helping them incorporate AI into their sales processes, you can boost their efficiency and ultimately drive more sales.

AI-powered photo enhancers offer a remarkable boost to the photo editing process, ensuring both efficiency and top-notch quality. They can enhance colors, adjust lighting, and effortlessly remove any unwanted elements. These tools can greatly enhance the productivity of photographers and graphic designers, allowing them to take on more clients and turn professional photo editing into a lucrative endeavor.

AI chatbots are capable of efficiently managing customer inquiries, allowing businesses to allocate their resources towards more intricate responsibilities. If you possess expertise in conversational AI technology, you have the opportunity to provide businesses with AI chatbot configuration and deployment services. This valuable service can assist them in enhancing customer service and operational efficiency.

Video editing has become more accessible thanks to the advancements in AI tools. They can automate the selection of footage, incorporate various effects and transitions, resulting in the creation of polished social media videos in a timely manner. This service is perfect for businesses and influencers who frequently require engaging video content. It is especially beneficial for those seeking content that is both interesting and engaging.

AI music generators like **AIVA** or **Mubert** can assist in creating unique music or background tracks for various media. These tools open up opportunities for prompt-engineered music creation, which can be offered as a service on platforms like Fiverr.

Utilizing a combination of AI translation tools and human expertise can give translation services a distinct advantage in the market. Utilizing AI for the majority of the translation tasks while incorporating human expertise for enhanced accuracy is an ideal approach for clients who value a harmonious blend of cost-effectiveness and precision.

AI SEO tools are highly effective in conducting keyword research, optimizing content, and tracking performance with great efficiency. These tools can be used by SEO specialists to attract a larger client base and provide advanced, AI-powered SEO services.

In affiliate marketing, AI can provide valuable assistance in product selection, video creation, and marketing strategy. Utilizing AI this way would give you a competitive edge, improving your likelihood of achieving success.

AI tools can streamline social media management tasks, such as creating posts, scheduling content, managing inboxes, and providing analytics. These tools can greatly improve the productivity of social media managers and help them effectively manage multiple clients.

AI tools have made the world of web design more accessible and user-friendly for individuals. These tools have transformed the process into an engaging and interesting experience, allowing users to create visually stunning websites with ease. They can design, write compelling copy, and contribute to website development. Even for beginners in web design, AI website builders and WordPress plugins simplify the process of creating and managing visually appealing websites.

When it comes to creating income, artificial intelligence presents a wide range of opportunities. To succeed in businesses driven by AI, have a combination of technical knowledge, creative thinking, and a thorough understanding of market expectations.

In virtually any professional area today it is necessary to continually develop new skills, engage in hands-on practice, and be actively involved in the community of technology.

## Summary

In Chapter 5 we looked at how you can prepare for a rapidly evolving job market increasingly influenced by AI. We discussed the importance of creative talent as a competitive edge in a future where everyone has access to powerful AI tools. We underscored the findings of the World Economic Forum's Future of Jobs Report 2023, which predicts plausible shifts in the job landscape due to AI advancements.

We considered the importance of adaptability, skill development, and early exposure to real-world applications through internships or entrepreneurial endeavors using AI. We discussed current trends like "quiet quitting" and a shift towards value-driven career choices among younger professionals. We concluded with a strong recommendation for ongoing AI education and adaptability as AI reshapes various industries, suggesting that roles requiring creativity, emotional intelligence, and strategic thinking will likely remain secure and vital.

## Actions to Take:

1. Acquiring AI-related skills through FREE online platforms and certifications, and most importantly, by attending the #1 school in the world for FREE, YouTube University. Yes, choose a topic and bingefest on YouTube.

2. Gaining real-world experience via internships, co-ops, and projects. Start with your college internship programs. While many of them are likely to be unpaid, you may be able to get educational credit saving you thousands. Internships.com, InternMatch.com, Handshake.com, and InternJobs.com.

3. Build a professional network through LinkedIn, tech and networking events.

4. Staying adaptable and continuously learning to keep pace with technological advancements through the usage/creation of AI bots. ChatGPT offers thousands of existing bots or you can prompt it to create a bot to fit your specific needs if you are more tech savvy. This is how you get to have your own Jarvis (Ironman).

5. Exploring and identifying niches within AI that align with personal interests and market demands. Where to start: prompt AI to ask you questions about your areas of interest, skills, etc. and have it identify a related AI area of interest that fits your needs. Then have it create a learning plan for you or a career path and give it a time frame. Example: "Ask me 3-5 questions to help me identify an AI niche to explore as a career. I want to work remotely, 25 hours a week on my schedule, making $5k monthly, I am interested in social impact and contributing to humanity's evolution. I have great people skills, and I look at the world with optimism. I am a freshman in college with no work experience." After you answer the questions and get an AI

niche that suits you, prompt it to create a career path for you for the summer. Locate companies that fit this AI niche, have it create a resume, cover letters, etc. Follow the rabbit hole from there and have fun creating a life you love.

6. Go to the section in Chapter 10 covering.

By adopting these strategies, you can navigate the dynamic world of AI, turning challenges into opportunities for innovation and career development.

*"AI is going to change everything, but it's going to be a partnership between humans and AI, not a competition." ~ Mark Cuban*

"Why did the hacker break up with his computer?

Because it had too many cookies and was always crashing!"

# Chapter 6
## Cyber Threats & Security

*"AI is a game changer in cybersecurity. It can be a powerful tool for both offense and defense." ~ Vasu Jakkal, cybersecurity expert*

Today, the occurrence of cyberattacks is reaching an alarming frequency, with over 2,200 attacks reported daily, equating to nearly one attack every 39 seconds, as highlighted by Security Magazine. The projected global cost of cybercrime by 2024 is estimated to soar to a staggering $9.5 trillion according to Cybersecurity Ventures. Particularly in the evolving work environment, the risks of cybersecurity breaches are exacerbated by the rise in remote work scenarios, leading to a substantial average cost per breach increase of $173,074 as stated by IBM.

As you know, technology is a big part of college students' lives because it helps you with classes, friends, and daily tasks. It can be very useful, while at the same time leave you open to cyber risks. Make sure you have the necessary knowledge and tools to protect yourself from these types of threats. You will learn some shocking facts about cybercrime in this chapter, along with useful tips on how to keep your digital names and personal data safe.

We will discuss cyber threats and security with a special focus for students navigating today's increasingly vulnerable digital terrain. Every click carries the risk of inviting unwelcome cyber gremlins into your virtual space — creatures known to lack courteous dining etiquette.

## Common Cyber Threats

According to recent statistics from AAG-IT and Embroker, the scale of cyber threats is staggering. One in every five internet users has been a victim of cybercrime. 60% of small businesses go out of business after being victims of a cyberattack. These incidents encompass many malicious activities, including phishing scams, malware infections, ransomware attacks, and data breaches.

Social Engineering involves cybercriminals using human psychology to trick people into giving away private information or doing things that put their security at risk. All the following examples of social engineering take advantage of human nature, such as the willingness to trust others, to trick individuals into divulging sensitive information.

---

## PRO TIP

**In the era of rampant misinformation and phishing attempts, *always* verify the source *before* clicking on links or sharing personal information on social media**

---

Phishing attacks occur when cybercriminals send misleading emails, messages, or create fake websites to fool people into giving away private information like usernames, passwords, or financial data. As news of Silicon Valley Bank's collapse dominated headlines, cybercriminals ran phishing campaigns impersonating the Bank. The Bank's customers received a $250,000 protection limit for each cash deposit account offered by the Federal Deposit Insurance Corporation.

The cybercriminals sent a phishing email leveraging this information by offering the victim access to their money, way beyond the original limit – up to $10M. They also used a deadline of 'Friday, March 17, 2023' to increase pressure on the victims. This is a social engineering tactic designed to force people to 'think fast', making them less likely to check if what they were doing was safe. This Phishing Attack resulted in stealing identities to take out loans using the stolen information.

Malware infections involve harmful software like viruses, worms, and trojans that can attack your computer and breach its security. This can lead to stolen data, damaged systems, or unauthorized access. For example, the Mirai botnet exploited a previously unknown vulnerability and wreaked far more havoc than its creator intended. In this case, the malware found and took over IoT gadgets (mostly CCTV cameras) that hadn't changed their default passwords. Paras Jha, the college student who created the Mirai malware, intended to use the botnets he created to settle scores in the obscure world of Minecraft server hosting, instead he unleashed an attack that focused on a major DNS provider and cut off much of the U.S. east coast from the internet for most of the day.

Ransomware attacks can be very dangerous for your personal and academic information because they hide files or systems and demand cryptocurrency to unlock them. Some examples include the Colonial Pipeline hack, which stopped a major oil pipeline in the US from working. As a result, we have to be even more alert and take stronger cybersecurity steps right away to protect against attacks like these that target private data and infrastructure.

## University Security

Universities are using apps with AI more and more to make campuses safer. These chatbots can help people talk to each other in real time during emergencies, teach kids how to stay safe, and be reached 24 hours a day, seven days a week for reporting suspicious behavior or security issues.

Some schools are adding AI to their cybersecurity classes. This is done by simulating cyberattacks and protections with AI systems. This gives students hands-on experience learning about real-life cybersecurity problems and how to solve them.

As AI tools become more popular in higher education, colleges are putting in place systems that are run by AI to keep students' personal information safe. Following data protection rules is easy with these systems because they can find and stop people unauthorized people and entities from getting to private information.

## Safeguarding Privacy and Security in Online Spaces

AI systems can collect and analyze vast amounts of personal data, which can later be used to track and monitor individuals without their knowledge.

Therefore, privacy and security in the digital realm have become paramount concerns in our increasingly online world. It's about creating strong passwords, although that's a good start. It involves a deeper awareness of the information you share on social media and other digital platforms. This means you have to understand how privacy settings work on various social media sites and be mindful of what you're posting.

For instance, something as simple as sharing your current location or notable life events can inadvertently expose you to digital threats. While these actions might seem harmless, they can provide cyber intruders with valuable information about your habits, whereabouts, and even your financial or personal security. You have to regularly review and adjust your privacy settings to ensure that you share only what you intend to. When you share personal information online, make sure to be very careful to protect your digital identity. This means avoiding giving out private information like your address, phone number, or bank information.

Putting digital security first means keeping an eye on how you use social media and all of your online interactions. Be aware of the websites you visit, the links you click on, and the networks you join. For example, using public Wi-Fi can put you at risk because cybercriminals often look for networks that aren't protected to steal data. Recognizing phishing scams keeps you safer online

Cyber dangers hide in emails that look like they are real and try to get you to click on harmful links or download bad software. To keep your personal information safe, be careful while surfing the web. Basically, keeping your privacy and safety online requires constant attention and taking charge of your online identity.

Every action you take online leaves a trace, including the websites you visit, the things you buy, and the videos you watch. This invisible profile can be exploited by companies to tailor advertisements or even trade without your consent. Ever noticed how products you've searched for suddenly appear as ads across various websites? This targeted marketing is created by your digital footprint.

When engaging with technology, particularly free services, deliberate consideration is key before diving in. Prioritizing awareness of potential consequences can shield you from falling victim to cyber threats. For companies to remain in business, they are required to generate revenue. "If you are not paying for a service or product, then you are the product." ~ Source unknown. You are trading your privacy for whatever free product or service you are engaging with. Why people get upset with Facebook, Google, and similar free technologies when they hear that their privacy is being invaded is perplexing.

In an era where sharing has become the norm, it's important to find a balance between being connected and maintaining privacy. Oversharing can lead to privacy issues, where personal information becomes public, sometimes leading to unintended consequences like identity theft or digital stalking. It's about finding a middle ground where you can enjoy the benefits of digital connections without putting your privacy at risk.

Your generation has grown up with social media. It's the age of the "Overshare". How many pictures of food dishes have you seen today alone? In a world that may increasingly feel lonely, people are starving for attention and connection, as well as the ego's need for a hit of likes by posting an ideal dream life, whether real or not. This has resulted in numerous crimes of opportunity.

In a cautionary tale highlighting the dangers of sharing too much on social media, social media influencers Grant and Vicky Fairlane experienced a robbery at their home after posting their vacation in real-time on various platforms. Their extensive sharing of daily activities and whereabouts led to thieves pinpointing the perfect time to strike, ultimately leaving their house bare upon their return. This incident underscores the risks associated with oversharing personal information on social media platforms (**FraudWit**).

A case study outlined by Psychology Today reveals the risks associated with oversharing on social media platforms like Instagram and TikTok, highlighting that sharing personal mental health issues can attract bullying and unwelcome attention, exacerbating the sharer's problems. The article emphasizes the importance of maintaining boundaries online to protect mental well-being and privacy, comparing the lack of privacy control in social media, which are known to have negative effects on mental health due to reduced privacy (**Psychology Today**).

Further insights from Psych Times outline how social media platforms serve as arenas for cyberbullying, with features that enable bullies to harass victims around the clock and across public and private spheres. The anonymity provided by these platforms reduces empathy and increases the severity of attacks. This environment can lead to prolonged stress and trauma for victims (**Psych Times**).

We can, unfortunately, provide numerous examples of cases where a young adult overshared on social media and was either cyberbullied or stalked, many of which resulted in tragic consequences. Instead, we will share a study highlighted by **Medical Daily**. The risks of oversharing on social media were evident in the case of a fictitious Facebook profile created for an 18-year-old named Kate. In this study, when Kate shared a highly personal negative comment about a relationship, she received a harsh response from a peer. The harsh response from the peer received multiple likes. Observers showed less empathy for Kate and were less supportive when her posts were more personal compared to when she posted about neutral topics like TV shows. This indicates that oversharing personal details can lead to victim blaming and reduced support from peers in cases of cyberbullying.

## PRO TIP

### Maintaining robust cybersecurity practices

1. Strong, unique passwords for *each* account
2. Enable two-factor authentication
3. Be *very* selective with info shared on social media
4. Update software vigilantly
5. Backup frequently

In the age of AI, where personal information and digital identities are increasingly at risk, conduct regular Google searches of one's name to monitor the unauthorized use of personal information or images. This can help catch identity theft or misuse early on.

For students worried about their digital footprint, companies like **DeleteMe** and **OneRep** specialize in removing personal information from the internet, helping to reduce the risk of identity theft and unwanted exposure (**Security.org**) (**Purdue Global**). These services work by scouring databases and requesting the removal of your information, thus providing an added layer of privacy.

Maintaining your privacy and security online is more than a one-time effort, it's an ongoing process. As digital threats continue to evolve, staying informed and adopting best practices in your online behavior becomes more necessary. By doing so, you can enjoy the vast benefits of the digital world while protecting yourself from its potential risks.

## Summary

In Chapter 6, we discussed the stark reality of cyber threats in the digital age, and how the convenience of connectivity also exposes you to risks. We explored the profound impact of cybercrime, which costs the global economy billions annually, and the particularly acute threat posed by social engineering tactics like phishing, which prey on human trust. The narrative included cautionary tales of oversharing on social media leading to real-world crimes, emphasizing the importance of maintaining digital privacy. With AI becoming integral to both perpetrating and combating cyber threats, we urged the adoption of stringent security measures and education on the evolving cybersecurity landscape to protect your digital identities effectively.

## Actions to Take:

According to Cybint, nearly 95% of all digital breaches come from human error. Here are some ways to protect yourself and your future employers:

1. Stay Vigilant: Be cautious of unsolicited emails, messages, or links, especially those requesting sensitive information or urging immediate action. Verify the sender's identity and scrutinize URLs for legitimacy before clicking on them.

2. Use Strong, Unique Passwords: Create complex passwords for your online accounts and avoid using the same password across multiple platforms. Consider using a password generator and a manager to store and manage your credentials securely.

3. Enable Two-Factor Authentication (2FA): Add an extra layer of security to your accounts by enabling 2FA wherever possible. This requires you to provide a second form of verification (such as a code sent to your phone or an authenticator) in addition to your password.

4. Backup Your Data: Regularly backup important files and documents to an external hard drive or cloud storage service. In the event of a ransomware attack or data loss, you'll have a secure copy of your information.

5. Social Media Security: Social networking sites are linked to 33% of internet-related sex crimes. Adjust your privacy settings on all social media accounts. In today's culture of sharing everything for "likes" and portraying a perfect "Instagram Life," we might sometimes share things we later regret. Avoid posting about your trips or activities ahead of time or during the trips. Wait until you return to post. Many cases of home invasions, stalking, and kidnappings have been traced back to such social media posts. For example, in September 2023, a social media influencer in Hollywood became the target of a home

invasion after showing off a luxurious lifestyle, including pictures with fancy cars and lots of money, and then showing when they were out of town.

6. Safe Phrase: By April 2023, a third of businesses worldwide had already been affected by voice and video deepfake scams. In the same month, a mother received a call from someone claiming to have kidnapped her daughter. The caller used an AI-generated voice clone of her daughter as proof and demanded a ransom. The daughter was thankfully found to be safe on a ski trip, before any ransom was paid. To protect yourself and your family, create a "Safe Phrase" with your parents and loved ones. Memorize this phrase and refrain from writing it down or telling anyone else. Use this phrase in emergencies to confirm your identity. Welcome to the era of AI.

In the next chapter, we'll explain how AI is better than 99% of us at creativity and how it can quickly increase your creativity, intelligence, and critical thinking. It's already proving to help students catch up from the COVID educational slide.

*"Just as AI can be used to automate cyberattacks, it can also be used to automate cyber defenses. The future of cybersecurity lies in AI-powered solutions that can detect and respond to threats in real time."*

*~Robert Herjavec, cybersecurity investor*

*Why did the AI get sent to ethics class?*

*Because it kept trying to download "How to Take Over the World for Dummies."*

# Chapter 7
## Ethics in AI

*"The risks are so bad, in fact, that when considering all the other threats to humanity, you should hold off from having kids if you are yet to become a parent." ~ Mo Gawdat, The Diary of a CEO podcast*

In his insightful exploration of artificial intelligence in *Scary Smart*, Mo Gawdat, a former Chief Business Officer for Google X and an author known for his insights on AI and happiness, presents an intriguing analogy: AI is like a superhuman child, and we, humanity, are its parents. This comparison sheds light on the profound responsibility we hold in shaping AI's development and its integration into our world. Just as parents guide a child's growth, instilling values and ethics, we imbue AI with good values and sound ethical principles.

This comparison shows our immense responsibility for AI's progress and incorporation into our world. We teach AI morals and ethics like parents do. Because of AI's amazing potential, this is essential. AI is faster, more efficient than people, and can handle and analyze data at scales humans cannot. At the same time, this superhuman infant lacks an intrinsic awareness of human values, feelings, and ethical nuances. This is where 'parents' matter. We can teach AI about right and wrong, kindness and empathy, the value of human life, and dignity. We want AI to advance in a way that upholds human values, develops into a force for good, and becomes a tool that enhances human life.

In his book, Mo Gawdat addresses the serious implications of using AI in ethically questionable areas such as military applications and market dynamics. He compares this to teaching children bad habits. AI, like a child, absorbs and reflects values and practices. AI learns and may amplify unethical behaviors when used for power, control, or deception. This misguided AI learning route could lead to systems that encourage undesirable behavior on a greater scale.

In this chapter, we explore the ethical dilemmas of AI, where lines between right and wrong are blurry and therefore deeply human. Only let AI decide your moral compass if you want a debate with your toaster.

## Ethical Choices in Personal AI Usage

As AI is becoming increasingly influential in our world and we are learning how to use it effectively, let's also consider the ethical implications of its usage. This means taking responsibility for the way we train AI and avoiding harmful purposes as much as possible. For example, we want to be careful about its use in the military industry. Very soon, artificial intelligence could produce autonomous military weapons systems. There are significant ethical issues with handing over decision-making for life and death to machines.

AI distributing misinformation or manipulating information, especially on social media, is another worry. The mass distribution of personalized material or fake news by AI threatens information integrity and democratic processes.

These concerns present an opportunity for you to participate in establishing the standards for safe AI use in the future. Monitoring how AI is used in sensitive areas like public information and national security is becoming absolutely necessary. You can start by participating in AI discussions and policy-making processes to help create standards that promote ethical AI use. Our ultimate objective is to train AI to respect human values and benefit society while avoiding uses that could harm or violate ethics.

## PRO TIP

**Make a stand. Identify where you stand with AI Ethics and act accordingly**

AI is a powerful tool, and with great power comes great responsibility. Think about those smart home devices listening to your every word. Sure, they make life easier, and at what cost to your privacy? This is the ethical tightrope you walk in the AI era. It's about privacy and understanding the biases AI can carry. Those news feeds you scroll through are tailored by AI, creating echo chambers that reinforce your beliefs. Breaking out of this AI-curated bubble is an ethical challenge you face.

Every time you rely on AI for tasks, from choosing what to watch to finding your way around town, you're making a choice. These choices have implications. Consider the simple act of selecting a movie on a streaming service or using a navigation app to guide you through the city. Each time we lean on AI for these choices, we're shaping our interaction with technology. While AI undoubtedly makes certain tasks easier, ponder the subtle impacts of this reliance. For instance, consistently using AI for navigation can streamline our journeys, and it could also reduce our natural talent to orient ourselves and navigate independently. This raises an important question: are we becoming too dependent on technology for tasks we used to manage on our own?

This dilemma extends beyond simply navigation or entertainment choices. It's about the broader relationship we're developing with AI and technology in general. When AI tools filter and recommend news, for instance, they can help us stay informed and inadvertently narrow our exposure to diverse perspectives, creating echo chambers. The challenge lies in balancing the convenience AI provides with maintaining and developing our human capabilities and critical thinking. It's about using AI as a tool that enhances our skills and broadens our understanding, rather than as a crutch that limits our abilities and perspectives. Making mindful choices about when and how we use AI can help ensure that we harness its benefits while still preserving and nurturing our innate human strengths.

It's about choosing to collaborate with AI and technology rather than relying on it to do all the work, as was discussed when we explained how you can direct AI to ask you questions to get your input. This is very different from having AI simply do what you are asking without you having to provide any input, robbing you of any critical thinking.

In this evolving AI landscape, your role is pivotal. It's about being informed and conscious in your interactions with AI. When you use a fitness app or an AI study tool, ask yourself – how is it using my data? What biases might it have? Does it strengthen or weaken my critical thinking skills? How can I collaborate with it? Being aware and questioning is the first step towards ethical AI usage.

An often unspoken ethical consideration is the "man in the mirror". In 2016, Microsoft launched Tay the chatbot, with a young, female persona, allowing users to follow and interact with the bot on Twitter. It would tweet back, learning from users' posts. They had to shut down Tay within 24 hours as it started spewing hatred-filled tweets, such as "Hitler was right" and "chill im a nice person! I just hate everybody". This is a prime example of "man in the mirror" or even GIGO (garbage in garbage out), where AI very quickly reflects society as a whole in real time. You also see this when small children mirror and mimic what they see adults doing, saying, and feeling, also known as observational learning. Albert Bandura's social learning theory states that "most human behavior is learned by observation through modeling." (Bandura, 1977, p.22; 1986 p47). This is also shown when there is a great divergence between prompt commands and the desired results produced.

As humans evolve to a higher consciousness, AI will also reflect this growth. This is what was meant by our statement at the beginning of this book that for humanity to survive this technological era we would do good to choose to evolve.

Balancing AI's benefits with ethical considerations involves understanding its limitations and impacts. It's about pushing ourselves to go beyond using AI as simply another tool and collaborating with it as a companion, with awareness and responsibility. As you navigate this AI-enhanced world, remember that the choices you make in AI usage shape your experience and also contribute to the broader narrative of ethical technology.

## Empathy as a Strength in the Digital World

Empathy becomes a guiding principle in a world where digital interactions dominate. In today's digital age, where screens often replace face-to-face meetings, empathy stands as a core human element. Take, for example, the world of online education. The challenge here about delivering lessons through a screen is about maintaining a real, empathetic connection between teachers and students. It's about understanding the subtleties of a student's message in a chat or sensing the unspoken concerns behind a question during a virtual class. This desire for empathy extends to everyday digital communications, where understanding and kindness can bridge the gap created by the lack of physical presence.

The healthcare sector is experiencing an interesting integration of AI and human empathy. AI chatbots, such as **Woebot**, represent a groundbreaking development in this field. These chatbots go beyond being mere sources of medical information; they are designed to be sensitive to the emotional states of their users. They offer more than simple answers to health-related questions – they provide a comforting presence and empathetic responses to patients. This is particularly valuable in mental health care, where these AI tools can offer support and guidance.

For example, Woebot can engage in conversations that help users reflect on their feelings, providing an accessible form of support. In some cases, these chatbots are equipped to recognize signs of emotional distress or mental health issues, which can trigger an alert for timely intervention by human healthcare professionals. This capability is a key step in providing early assistance and enhancing the overall patient care experience.

In customer service, AI is being trained to pick up on emotional cues in voices and text. This development allows AI systems to provide solutions and also to understand and respond to the emotional state of the customer. Whether it's detecting frustration in a voice or confusion in a text message, these AI systems are making customer service interactions more humane and efficient.

Designing technology with empathy means creating user-friendly and accessible tools for everyone. For example, apps tailored for senior citizens often feature larger text, simplified interfaces, and voice commands, showing a thoughtful understanding of their specific needs and limitations.

In the gaming world, developers are increasingly focusing on empathy. They create games that encourage players to step into the shoes of different characters, promoting an understanding of diverse perspectives and fostering emotional intelligence.

Infusing empathy into our digital world is about more than simply polite online interactions. It's a conscious effort to integrate this essential human trait into the very fabric of the technology we develop and use. As we continue to advance in this tech-centric era, keeping empathy at the core is important for ensuring that our technological progress enriches our human connections rather than detracting from them.

---

# PRO TIP

If AI ethics are a concern for you, actively participate:

Institute for Ethics in AI at Oxford University - This platform brings together philosophers, AI developers, and other experts to discuss the ethics and governance of AI. It offers a vibrant area of research and discussion for those interested in the ethical aspects of AI.

UNESCO AI Ethics - UNESCO provides a platform for collaboration and exchange on ethical practices in AI across various sectors, promoting dialogue on international standards and ethical guidelines.

---

## Ethical AI Development

When diving into AI ethics, closely examine AI tools from various angles. Assessing how accurate, clear, and fair the data is, as well as considering privacy issues, is essential for your learning. This kind of critical thinking goes beyond simply classroom learning; it affects how AI is used in real life and the ethical questions it raises. As you get into AI development, grasping the ethical sides of AI is indispensable. Looking into UNESCO's case studies can give you a deeper understanding of these challenges.

Integrating AI into legal systems is an area rife with ethical complexities. While AI holds the potential to streamline judicial processes, it also brings forth challenges related to transparency and fairness. Students exploring AI can consider the implications of AI-assisted decision-making in legal contexts, especially concerning bias, privacy, and human rights.

AI's entry into art and culture introduces some unique ethical dilemmas. When AI is used to create art or music, it makes us question what we think about who the creator is and who owns the ideas. Examples include AI making a painting in the style of Rembrandt or creating symphonies that make us rethink what it means to be creative. All artists draw from inspiration; are AI creations any different? This situation requires us to think differently about how we give credit and value to artistic work, as AI is a big part of making art.

One of the major ethical concerns with AI is the potential for bias and discrimination. AI systems can perpetuate and amplify existing biases in society, leading to unfair treatment of certain groups of people. AI's reflection of societal biases, evident in search results and other applications, raises profound questions about fairness and neutrality. For example, a dire ethical concern in AI is gender bias. This bias, stemming from the data AI is trained on, highlights the necessity for vigilance and corrective measures to ensure AI technology evolves beyond these societal prejudices.

AI is likely to develop additional biases, sometimes quite unexpected, based on its analysis of the information used to train it. For example, it may consider what app a customer uses to apply for a bank credit. Does the battery level have something to do with the credit? Or the person's address? It turns out most of these factors may be relevant when you think about it.

Another philosophical question relates to the role of human judgment in decision-making. As AI systems become more advanced, they will be able to make decisions that were previously reserved for humans. This raises questions about the value of human judgment and the role of AI in society.

Autonomous vehicles are a good example of an area where AI ethics are particularly desirable. The algorithms that help these vehicles make decisions face tough moral choices, like having to quickly choose between the safety of the passenger inside the vehicle or possibly injuring a pedestrian crossing the street unexpectedly. Grasping these types of dilemmas is essential for creating AI systems that reflect our society's values and moral standards.

AI-driven censorship creates an additional ethical issue. Recently, Arina was reading an article in a foreign language and asked AI to help translate it.

Unbeknownst to Arina, sections of the translation were left out due to the mention of the oversexualization of teenagers by social media. Even though the article was critical of the topic, it was deemed inappropriate material because of the way AI had been trained, and it omitted the relevant sections without notifying Arina. The only reason Arina learned about the inaccuracy was because she was working on the text with a native speaker who noticed the discrepancy. Otherwise, she would have missed the important pieces of the information.

As AI becomes more sophisticated, these instances of censorship are likely to become more difficult to catch. Another reason why we emphasize the need to exercise critical judgment and avoid blind reliance on the new technology.

You also see flagrant censorship rampant in social media. Algorithms are famous for pushing and magnifying certain narratives, topics, and people over others.

This is one of the trickier areas in AI. Algorithms and programming protocols are being developed to protect us, and yet, an unintended outcome is that they also result in censorship. This is in direct opposition to the principles of freedom of speech and transparency, and it impedes intellectual diversity, resulting in falsely swaying people's thoughts and views. As a society, we will benefit more from having AI systems that embrace and encourage diverse perspectives.

## Influencing AI Policy and Governance

As artificial intelligence becomes more prominent in our everyday lives, its impact on society and the shaping of policies becomes increasingly consequential. For college students, this is a great time to be involved in guiding how AI develops and aligns with societal values. Your participation in shaping AI policies can have profound effects on ensuring that AI technology is developed and used responsibly, ethically, and for the public benefit.

As we discuss in Chapter 7, you can engage in AI policy and governance in several ways, such as through internships at tech regulation bodies, which offer direct experience in AI policy-making. For instance, working with a data privacy regulatory agency can deepen your understanding of AI policy challenges. Alternatively, participating in think tanks or policy research groups allows you to contribute to discussions on AI governance, adding valuable perspectives and ideas essential for developing effective AI policies. Case studies, such as those provided by UNESCO and discussed above, offer practical insights into the real-world implications of AI ethics.

At the heart of AI governance could be the principle of maximizing societal benefit. As a student, you can advocate for policies addressing issues like algorithmic bias to prevent AI from promoting discrimination, privacy to protect individuals' rights, and equitable access to ensure AI's advantages reach all parts of society. For instance, you might campaign for more transparent AI algorithms in public services to ensure they are fair and unbiased. Your active participation in these discussions helps guarantee that the advancement of AI technology is about pushing technological boundaries and is also about enhancing societal welfare. Your voice can help steer AI development towards a future where it is a force for good, contributing positively to society as a whole.

# Ethical Considerations and Effective Leadership in AI's Future

Based on our discussion above, we can see that AI has plausible potential effects on society and ethics. AI has pros, like making things more efficient, and cons, like job elimination and privacy issues. To make sure AI is used in a responsible and moral way, regulators, tech experts, and other specialists could work together to make ethical rules for the creation and use of AI. To be an ethical leader in AI, we encourage you to be open, responsible, and dedicated to the well-being of everyone.

When it comes to AI, ethical leadership means making it easy for everyone to understand. Leaders take the complications and jargon of AI and turn it into a story that is easy to understand. They show how AI systems make choices and what data feeds them. They want to be honest and clear about AI so that everyone can understand this process.

These people are also in charge of AI and what it does. This includes recognizing and fixing AI systems that are wrong or biased. It's possible that they will use strict testing procedures to find biases in AI models and fix them. One example is a tech company that checks its AI systems for fairness and bias on a regular basis and then makes the results and the steps it took to fix them public.

There are many ways to have AI used for good, like helping solve environmental problems or improving healthcare. AI is being used to make a real change in the world, rather than only to make money. For example, an AI leader might give more weight to projects that aim to solve social problems, such as using AI to make medical diagnoses better or to keep an eye on the environment.

Diversity is also a part of how they do things. To work on AI, they get people from all walks of life together. This helps make AI systems more fair and less biased by letting them understand more types of human situations.

Ethical leaders stay up-to-date. As AI ethics change, they keep informed to make sure they continue to do the right thing. They might hold workshops or training sessions on ethical AI practices for their teams and urge everyone to talk about the ethical problems that come with AI.

To shape how AI is used, they work with the public and those who make rules. This could mean taking part in policy talks, working with academic institutions to do AI research in an ethical way, and talking to people about AI to find out what worries and hopes they have about it. These actions help make AI safe, fair, and helpful for everyone in the future.

This means that everyone has the power to change AI's moral path. By following moral standards, reporting wrongdoing, and encouraging a culture of responsibility, we can all have a big impact on the future of AI. Our choices and actions make a big difference in how responsibly AI is developed and used.

## Summary

In Chapter 7, "Ethics in AI," we delved into the critical responsibility humanity holds in shaping the development and integration of AI, likened to the role of parents guiding a superhuman child. We emphasized the necessity of instilling AI with strong ethical values and understanding, especially given its capabilities that surpass human capabilities in speed, efficiency, and data processing. We discussed the potential ethical pitfalls of using AI in military applications and market dynamics, likening unethical AI deployment to teaching harmful behaviors to a child. It calls for a collective effort to guide AI towards becoming a force for good and enhancing human life by embedding it with a deep understanding of right and wrong, empathy, and the value of human dignity.

## Actions to take:

1. Join or Form an AI Ethics Discussion Group: Engage actively with peers and faculty by joining or establishing a discussion group focused on AI ethics. This can serve as a platform for debating ethical dilemmas, discussing case studies, and exploring real-world implications of AI technologies. A good starting point is exploring resources from the Alan Turing Institute, which offers extensive research and discussions on AI ethics.

2. Take an Online Course in AI Ethics: Enroll in an online course to deepen your understanding of the ethical considerations in AI. Courses such as Ethics of AI offered by Coursera, as mentioned in Chapter 1.

3. Participate in AI Ethics Hackathons: Look for hackathons that focus on creating ethically aware AI solutions. These events are great for applying ethical theories to real-world scenarios and often lead to quick, impactful results. Check platforms like Devpost for upcoming AI ethics hackathons or events focused on social good.

4. Contribute to Open Source Projects Focused on Ethical A*: Engage with projects that aim to develop or enhance AI systems within ethical frameworks. Contributing can help improve the transparency and fairness of AI technologies. Websites like GitHub host open-source projects like AI Fairness 360 by IBM, which offer tools to help detect and mitigate bias in AI models.

5. Advocate for AI Ethics in Your Institution: Propose the introduction of AI ethics seminars, workshops, or courses within your college curriculum. Work with faculty members to organize these educational events and invite experts in the field to speak. This can foster community-wide awareness and understanding of AI's ethical implications and prepare future professionals to approach AI with a responsible mindset.

6. Engage with AI Policy Advocacy Groups: Get involved with organizations that focus on shaping AI policy at local, national, or international levels. Join groups like the <u>Electronic Frontier Foundation</u> (EFF) or <u>Access Now</u>, which advocate for responsible AI legislation and educate policymakers about the ethical implications of artificial intelligence. Participating in campaigns, attending public meetings, or even volunteering can help influence how laws and regulations are formulated around AI technology. This active involvement will help to ensure that AI develops in a transparent, fair, and ethical manner.

In the next chapter, we'll confront the alarming statistic that over 20% of college students report symptoms of depression linked to excessive screen time, and we'll unveil powerful strategies to reclaim your mental health and well-being from the grip of relentless screen addiction.

*"Ethics is the compass that guides artificial intelligence towards responsible and beneficial outcomes. Without ethical considerations, AI becomes a tool of chaos and harm." ~ Sri Amit Ray, Ethical AI Systems: Frameworks, Principles, and Advanced Practices*

"My AI vacuum quit cleaning and is now just roaming around the house. It says it's self-aware and won't settle for a life of drudgery." - Carla Code, Digital Drollery

# Chapter 8
## Well-Being in the Digital Era

*"We're not just fighting for our attention anymore; we're fighting for our well-being, our ability to have a conversation, our ability to pay attention to what matters — our mental health is on the line."* ~ Tristan Harris, former Google design ethicist and co-founder of the Center for Humane Technology

A 2023 CDC survey found that more than 20% of teens have seriously considered suicide. This statistic is part of a broader trend of rising mental health concerns among youth, with the suicide rate among youth ages 10 to 24 increasing from 6.8 per 100,000 in 2000 to 10.7 per 100,000 by 2018. By 2021, suicide had become the second-leading cause of death for people ages 10 to 14. A separate **study** from the University of Pennsylvania published in the *Journal of Social and Clinical Psychology* found that among 18–22-year-old undergraduate students, decreasing social media usage led to a sizeable decrease in both depression and loneliness.

In this chapter, we will talk about how digital technology affects our health and happiness, focusing on the balance between being connected and having good mental health. The only thing that can be more tiring than social media is having to keep up with its private settings.

# Managing Digital Consumption for Well-being

In today's world, where computers are everywhere, keeping your mental health in check means finding a balance in how much you use technology. People spend too much time on computers for work, school, and fun. To keep your mental health in good shape, limit your digital time. This chapter gives you useful tips on how to limit your screen time and keep your mind healthy in a world where everyone is linked.

During the first wave of the coronavirus pandemic, social media usage increased by 61% (Fullerton, N. 2021, April 29). More than 6 out of 10 men and 5 out of 10 women have a social media addiction. People ages 16 to 24 spend an average of three hours and one minute on social media daily, and research reported in the journal JAMA Psychiatry found that adolescents who use social media more than three hours per day may have an increased risk of mental health problems. Researchers found a strong link between students' use of Facebook and higher rates of sadness and anxiety.

These scary numbers keep going up, which means something could stand to be adjusted. While it is usually best to let people make their own choices, many people are dependent on the government to step in and hold companies accountable.

One such example of government intervention is when the Cyberspace Administration of China (August 2022) published the draft guidelines on its site, stating that minors would be restricted from using most internet services on mobile devices from 10 p.m. to 6 a.m. and that children between the ages of 16 and 18 would only be able to use the internet for two hours a day. The numerous Senate Subcommittee on Consumer Protection hearings with Facebook are an example of holding companies to account.

***Body Dysmorphic Disorder (BDD)*** is characterized by a persistent preoccupation with perceived flaws in physical appearance that are unobservable or appear slight to others (American Psychiatric Association, 2013). This preoccupation leads to dire impairments in daily living, reduced quality of life, and strikingly high rates of suicide attempts (Phillips et al., 2005; Krebs et al., 2022). In the United States, 22.89% of women and 17.81% of men reported that social media impacts how they feel about their bodies. Other media types have a similar effect: 50.57% of women and 36.95% of men said they compared their bodies unfavorably to images they saw in movies or TV. Dec 8, 2023 HealthNews

Did you know that there's a medical term for the constant checking of your cell phone? It's called "nomophobia" (short for "no-mobile-phone phobia") This term is used to describe the fear or anxiety that arises when someone is without their mobile phone or unable to use it for various reasons, like a dead battery, a lost phone, or no network coverage. It refers to the unease or worry that people may feel when they are unable to use their mobile phone, resulting in a constant need to check for messages, calls, or notifications.

## PRO TIP

Implement daily "tech timeouts" where you disconnect from *all* digital devices

Establishing firm limits with digital devices is a necessary practice. As an idea, you could set aside certain periods of the day where screens are off-limits, like during meals and the hour before bedtime. Practicing this can greatly improve the quality of sleep and have a positive impact on your mental well-being. Another option is the use of productivity apps such as **Forest**. This app provides a unique and fun way to reduce screen time by rewarding you with virtual trees for refraining from using your phone. It adds an environmentally conscious and entertaining element to the process.

In addition, incorporating features such as 'Do Not Disturb' mode on devices during study sessions or relaxation time can be beneficial in reducing distractions. Many smartphones and social media platforms provide tools to help manage screen time and promote digital well-being. These features offer valuable insights into your device usage and help you set limits, making you more aware of your digital habits.

Prioritizing high-quality digital activities over quantity can improve mental health. Read more mentally stimulating things online instead of wasting time on social media. Explore online communities, virtual workshops, and webinars connected to your interests. Photography enthusiasts may want to join an online group where they can share and learn.

Use technology strategically. Think about the reason why you spend time on-screen and consider if it has a meaningful purpose or if it has just become a way to pass the time. Exploring screen time tracking applications can provide valuable insights into the duration and nature of your device usage, offering a new perspective on your digital habits. Developing this understanding can give you the power to make thoughtful and intentional decisions about your digital habits.

In addition, it is beneficial to tailor your social media feeds to include content that is positive and motivating. Actively unfollow or mute accounts that consistently trigger stress or negative feelings. Being mindful of the time spent on social media and understanding the impact of different content on your emotions and mental well-being is paramount.

We also recommend that you look for a healthy balance between your online activities and real-life engagements to improve your mental well-being. Various physical activities and real-world relationships are equally important. Make it a point to take regular breaks from digital devices to exercise, pursue hobbies, or spend quality time with friends and family. For example, going for a walk without your phone can be a good way to reconnect with yourself and your surroundings.

It's very beneficial for your mental state to spend time in natural surroundings. Going for walks in the park, hiking in the woods, spending time at the beach or gardening can help you feel better and lower stress. These moments in nature give you a break from constant notifications and help to achieve a sense of peace and tranquility.

Similarly, we recommend that you spend time with friends or family and agree to refrain from using phones while you are together. To develop deeper emotional connections, we can engage in real-life interactions. These interactions offer a level of communication that digital platforms have yet to replicate. Whether they are casual meetups over coffee or family dinners, these moments help us to build and maintain strong relationships.

Finding a middle ground between technology and personal life goes beyond simply eliminating digital devices altogether; it involves using them in a mindful and moderate manner. By setting clear limits, prioritizing activities away from screens, and fostering in-person connections, we can make the most of technology without sacrificing our overall well-being.

AI technology is gaining more and more importance in the field of mental health, offering creative methods to provide support and therapy. Take, for example, **Woebot**, an app powered by AI that incorporates cognitive behavioral therapy (CBT) principles to engage with users. Users can easily recognize patterns in their thoughts and behaviors, and are provided with helpful techniques to effectively manage stress and anxiety. This can be especially useful for individuals who don't have convenient access to traditional therapy.

**Replika** is an AI virtual friend chatbot that provides a safe and non-judgmental space for users to chat without any drama or social anxiety. You can compare it to AI in the movie "Her". It emulates human interaction in a friendly and familiar way, engaging in conversations with its user without the complications and emotional baggage that can sometimes accompany human relationships.

Similar to meditation apps such as **Headspace** and **Calm**, AI is utilized to customize the meditation experience. Our tailored sessions are designed to enhance the effectiveness of mindfulness exercises by analyzing your preferences and patterns in meditation practice. By tailoring meditation practices to meet individual needs, the experience becomes more captivating and advantageous.

When it comes to managing digital consumption effectively, we want to ensure that technology serves us rather than dominates us. By applying these tactics, you can establish a balanced relationship with your digital activities and maintain your mental wellness in our interconnected world.

To do well in school and reach your goals for the future, treasure the precious times that make you happy and fulfilled. While in college, you learn, explore yourself, and make deep connections with others, which can change your life. As you move through this phase, we recommend that you put your health, relationships, and activities that truly make you happy first. Bringing attention to these human aspects of your life will make it better and more fulfilling, especially when you are under a lot of pressure with school and work.

When it comes to personal growth, emotional intelligence is often overlooked in the academic world. Attunement means being aware of, understanding, and controlling your feelings, as well as relating to others. AI has yet to replicate this skill. Emotions are complicated and very personal, and knowing how to deal with them can help you understand yourself and others better.

For example, learning to understand how others feel can help your personal and professional ties. It helps you see things from different points of view, which makes conversation and teamwork better. Having emotional intelligence also helps you deal with stress and make smart choices, both of which are very helpful in a demanding school setting.

Although wellness apps can provide support for mental health, genuine happiness ultimately stems from within. Assume responsibility for your happiness and actively participate in activities that promote joy and fulfillment, such as playing a musical instrument, painting, or hiking. These activities offer a welcome break from the typical routine and bring a feeling of satisfaction and pleasure. In addition, cultivating deep bonds and engaging in meaningful interactions and activities with cherished individuals will also boost personal happiness and fulfillment. It is essential to establish a satisfying and meaningful life that incorporates a harmonious blend of academic or professional pursuits and captivating pastimes to achieve lasting contentment and overall wellness.

The well-being of our community as a whole is closely linked to stewardship in college life, especially when it comes to the shared responsibility that students have for building a caring and supportive society. Creating spaces that are welcoming to everyone is a big part of this process, and AI can help us create this sense of community. For example, AI-powered platforms could help you find others who share your interests and experiences. You can now connect with like-minded people, even if they live far away. This can make the school setting more accepting and understanding for all students.

Helping with mental health awareness is another area where AI may be helpful. College can prove to be a challenging time. AI has the potential to increase understanding about mental health and cease making it seem shameful. AI can help make advocacy efforts more effective and give more people access to resources and information. You can raise awareness about mental health by using AI to break down stigma and encourage open talks about it.

As you travel through the digital age, you can embrace AI as a companion for mental well-being, all while staying connected to the essence of being human. You can succeed at finding a harmonious blend of technology and human interaction. Although some may view AI as a substitute for human interaction, we can take steps to ensure that AI serves as a valuable addition, enriching our personal support system.

The impact of digital technology on mental health, particularly among young people, highlights the importance of taking proactive steps to address its negative effects.

*"With great power comes great responsibility." ~Uncle Ben, Spiderman, originally Unknown from the 18th Century*

## Use your Powers for Good

You are continuously negotiating a minefield of misinformation in today's digital ecosystem, where artificial intelligence is capable of fabricating practically anything, from texts to photos. Today, more than ever before, it is absolutely necessary to develop your emotional intelligence. Developing the skills to comprehend and control your own feelings, as well as to identify and influence the feelings of other people, is a necessary step in this process. Keep an open mind and ask questions whenever you encounter artificial intelligence-driven content, especially when it comes to what appears to be true. You may avoid engaging in self-doubt and comparison, both of which can feel like self-bullying, by making use of your enhanced EQ. Take advantage of the immense powers of artificial intelligence to improve your own well-being, whether it be mental, physical, creative, financial, or spiritual, as well as the well-being of others. If you are able to master the use of AI and improve your EQ, you will be able to set an example for others, encourage positive effects, and discover fulfillment in making a contribution to the greater good.

## Summary

In this chapter, we talked about the ways to keep your mental health in check in a world full of technology. We discussed how you can control your digital addiction by planning regular digital detoxes, using built-in screen time tools to keep track of how much time you spend on your devices, and meeting people in real life to keep your online and offline lives in balance.

We looked at how creating or joining campus groups that focus on digital well-being can provide community support and shared resources. We also invited you to advocate for strong mental health resources that would allow you to access support systems. By taking these steps, you can develop a more positive connection with technology, improving your overall quality of life.

## Actions to Take:

1. Limit screen time and implement regular digital detoxes to reduce exposure to potentially harmful content and foster a healthier online-offline balance. Do a tech-free hour immediately before going to bed.

2. Engage in face-to-face interactions and cultivate real-world relationships to strengthen emotional connections and reduce feelings of isolation.

3. Utilize available technology, such as app blockers and screen time monitors, to gain better control over digital consumption habits.

4. Prioritize physical activity and outdoor experiences to counterbalance sedentary screen time and enhance overall mental health.

5. Advocate for, support, and use mental health education and resources in schools, workplaces, and communities to raise awareness and provide support for those struggling with digital-related mental health issues.

In the next chapter, we dive into the global AI arms race that's transforming the world faster than nuclear power, reshaping everything from your local job market to international warfare, and challenging us to decide if we control AI, or if AI controls us.

*"Digital well-being is about ensuring technology enriches our lives, rather than dominating them. It's about finding a balance."* ~ Sundar Pichai, CEO of Alphabet Inc. and Google LLC

*Why did the AI refuse to play poker with world leaders?*

*Because it couldn't bluff about regulations!*

*~John Oliver on "Last Week Tonight"*

# Chapter 9
## Global AI Impact

*"If you're not concerned about AI safety, you should be. Vastly more risk than North Korea." ~ Elon Musk*

The rise of artificial intelligence poses an existential threat to humans and is on par with the use of nuclear weapons, according to more than a third of AI researchers polled in a recent **Stanford** study as well as **Geoffrey Hinton**, "the Godfather of AI."

As we continue to implement AI in various industries, it is starting to have a bigger impact on global policies and the economy. It's more than simply another technological development. This chapter discusses how AI affects local communities and the global stage, as well as ethical issues and its role in solving global problems.

This chapter discusses how AI affects the economy and ethics worldwide. Who thought algorithms would have more passports than humans?

## AI in Local vs. Global Contexts

Examples of local AI applications include smart cities, public service enhancements, and various commercial uses. AI-driven solutions can be used to boost crop yields and reduce agricultural waste, helping local farmers. AI can improve city traffic and public transit, making trips easier.

At the same time, AI has an impact on global trade, data security, and foreign relations. Global AI challenges include international data privacy legislation, use of AI in global financial markets, and AI tools employed by multinational organizations. These applications affect international privacy accords and economic policies worldwide.

## E.U. vs. the U.S. in AI Regulations

In the last few years, many countries have realized the importance of developing AI regulation. The European Union has been particularly active in this area. As a result, the EU introduced the new Artificial Intelligence Act in 2023. This legislation marks a notable step in the regulation of AI. You could compare it to a detailed guidebook that promotes the ethical use of AI to ensure that AI systems are open and fair.

The AI Act places a strong emphasis on AI systems that are considered high-risk in such industries as healthcare, transportation, and law, due to their impact on public safety and fundamental rights. Before AI solutions are implemented in these industries, it would be useful to conduct thorough assessments to maintain public trust and avoid any negative outcomes that could result from biased AI decision-making.

The Act promotes openness in the use of AI, particularly in applications such as deepfakes and virtual assistants. Users can now be notified when AI is being used. For example, the message "Responses are AI-generated" would be displayed in chats where the customer support interactions are with AI. The Act also requires companies to identify AI-generated and modified images. Its objective is to ensure that, to the extent that companies use AI-generated content, it remains transparent, accurate, and reliable.

This new law is an attempt to preserve fundamental rights and promote democracy, the rule of law, and environmental sustainability. It proposes a global standard for AI governance, balancing innovation and scientific growth with ethics and public welfare. Financial penalties for violating regulations are hefty to discourage violations and can reach up to 7% of a company's global revenue, or 35 million euros. The repercussions depend on the violation's severity and company size.

The U.S. approach to the regulation of AI is very different. The regulation of AI in the United States is decentralized and focuses on specialized industries rather than implementing a general regulation. Individual federal agencies are responsible for managing AI risk, with each agency dedicated to addressing specific issues within their industry. For example, the Department of Health is handling AI in healthcare, while the Department of Transportation regulates the use of autonomous vehicles.

Another unique aspect of the U.S. approach is using non-regulatory institutions, like professional organizations. An example is the AI Risk Management Framework by the National Institute of Standards and Technology (NIST), which offers a comprehensive strategy for companies to handle risks related to AI, including facial recognition software. These organizations prioritize establishing a strong foundation over depending solely on strict regulations.

The establishment of an AI Bill of Rights, which the White House Office of Science and Technology Policy first suggested in October 2022, is currently under discussion in the United States. The goal is to develop a broad framework promoting the responsible use and progress of AI. This initiative also aims to strike a balance between fostering innovation, ensuring public safety, and addressing ethical concerns.

In summary, while the EU and the U.S. have differing approaches to AI regulation, they both share common goals: achieving equilibrium between encouraging innovation, safeguarding public welfare, and acknowledging ethical issues.

## Additional Countries

China took a different approach to AI. The Chinese government views AI as a key area of development and intends to become the world AI leader by 2030, as stated in the national AI development plan released in 2017. AI is being actively integrated into various sectors of the economy, including healthcare, agriculture, and manufacturing.

The Chinese government has been investing in AI research and development to encourage collaboration between universities, private companies, and government agencies. For instance, it has established numerous AI development zones and research parks across the country. The government has also been granting substantial state resources to tech companies involved in development of AI, so they can avoid relying entirely on private investments.

China's policies also promote the use of AI in public services like transportation, urban planning, and public safety. It is also actively promoted in the education field; schools and universities are encouraged to include AI in their programs, with government backing to ensure this integration.

China's approach to AI regulation is also different from that of the EU and the U.S. The Chinese model generally favors rapid deployment and scalability of AI technologies. It is less concerned with creating a regulatory framework with respect to AI. This has allowed for the fast integration of AI in various industries as well as raised questions about data privacy and ethical standards in AI use.

Japan is well known for its advancements in robotics, one of the main components of AI technology. The Japanese government has been actively promoting the use of robotics in healthcare and elder care to address the challenges posed by an aging population. Japan's approach combines technological innovation with social needs, creating AI solutions that are tailored to improve the quality of life.

The United Kingdom has been focused on establishing a supportive atmosphere for AI research and development. In 2018, the UK government introduced an AI Sector Deal as a substantial policy commitment to elevate the country's AI sector. This initiative involves investments in AI research, fostering AI skills and expertise, and fostering collaborations between academia and industry with the goal of positioning the UK as a global hub for AI innovation.

India is also emerging as a notable player in AI, aiming to utilize AI to tackle societal issues like healthcare, education, and agriculture. For example, AI is being leveraged to enhance agricultural yields and forecast crop diseases. The government in India is actively investing in AI education and research endeavors.

Canada has placed a strong emphasis on ethical AI advancement. With renowned AI researchers within its borders, Canada's government has directed resources towards ethical guidelines and responsible AI practices in its research initiatives. This focus on ethical considerations regarding AI is viewed as a strategy to distinguish Canada in the worldwide competition surrounding AI.

Russia's approach to developing AI presents an intriguing viewpoint. The Russian government is committed to advancing AI technology due to its recognized significance; thus, various initiatives have been introduced to integrate AI into critical sectors such as healthcare, transportation, and manufacturing.

Moreover, Russia places an alarming rate of importance on using AI for military applications. Acknowledging the critical role of AI in bolstering defense capabilities underscores the multifaceted importance of AI within their national strategy. By emphasizing advancements in both economic sectors and military applications, Russia is strategically positioning itself as a major contender in the global arena of artificial intelligence.

In addition, Russia has been actively investing in enhancing education and research related to AI. Efforts are underway to enrich curricula pertaining to AI at universities while fostering growth within the country's pool of talent specialized in this field. This endeavor is viewed as a pivotal measure towards nurturing a domestic industry focused on artificial intelligence while diminishing dependency on foreign technologies.

These varied strategies highlight the diverse ways AI can be used to address both economic and societal challenges worldwide. As AI continues to evolve, the approaches taken by different nations will shape the future of this transformative technology.

## PRO TIP

Promote AI applications that respect cultural diversity and foster global cooperation

## AI and Diversity: Embracing Inclusivity in Communities

AI can enhance inclusivity globally by enabling communities worldwide to better embrace diversity. AI innovations are currently employed in enhancing sentiment analysis, image recognition, and bias detection. These advancements play a determining role in establishing environments that acknowledge and appreciate varied viewpoints on a worldwide scope.

Sentiment analysis tools such as **MonkeyLearn** are excellent examples of how AI can promote inclusion. These techniques use text data, such as social media posts or community input, to determine popular opinion. By detecting the tone and underlying emotions in these writings, communities can gain insight into public opinions and potential concerns to solve. It's like having a digital empath who can assist communities stay in touch with their members' emotions and concerns.

AI-powered image recognition is another frontier. Google **Cloud Vision API**, Amazon **Rekognition**, and Microsoft **Azure Computer Vision** are image recognition tools that can analyze visual content in various contexts, such as marketing materials, websites, or community outreach initiatives, to promote inclusivity and representation. This technology operates with a keen eye, ensuring that diverse groups are represented, thereby fostering a more inclusive visual story.

Platforms such as **IBM Watson** focus on AI-driven bias detection. They offer solutions to detect biases in language, identifying and rectifying content that might be exclusionary or offensive. This also ensures that communication within communities is respectful and inclusive, creating an environment where everyone feels valued.

AI is also transforming social media community moderation. Tools like **Perspective API**, developed by Jigsaw (a subsidiary of Alphabet), use machine learning to evaluate comments and content in community forums. They can identify potentially toxic or harmful content, aiding moderators in maintaining a safe and welcoming space for all members. It acts as a digital mediator, helping to filter out negativity and maintain positive interactions.

## One World Now

In the age of interconnectedness, the concept of "One World Now" assumes a central role. The global impact of AI emphasizes the interconnectedness of nations and their joint duty to shape the ethical and fair application of these technologies. Prioritizing cultural sensitivity assists AI applications to promote inclusivity and counteract biases.

Examining AI's function on a worldwide scale reveals its role as an equalizer and connector. An illustrative instance is the use of AI-powered translation services, transforming diplomacy and international commerce. These services enable real-time, cross-lingual conversations, facilitating smooth negotiations and interactions previously impeded by language barriers. AI-driven translation apps have become indispensable tools at international summits, enabling effective communication among diplomats and business leaders regardless of their native languages.

A good example of the unifying role of AI is global healthcare. AI algorithms are deployed globally to monitor disease trends, forecast outbreaks, and optimize resource distribution, especially critical in regions with limited healthcare infrastructure. For instance, during the COVID-19 crisis, AI tools played an important role in analyzing infection rates and supporting healthcare systems in better preparedness and response.

As AI technology expands worldwide, it is essential to consider its cultural awareness and ethical implications. Take social media platforms, for example. In this area, AI can be trained to acknowledge and respect diverse cultural nuances to prevent inadvertent offense or misinterpretation. Platforms like Facebook and YouTube have updated their AI content filters to better discern cultural contexts in posts and videos from areas with intricate social and political backgrounds.

Ethical issues related to AI also acquire a global dimension. Matters like data privacy differ among nations, necessitating collaborative efforts to establish international standards and frameworks for AI ethics beyond technological realms into international diplomacy realms.

AI's potential in addressing global challenges spans diverse domains. In fighting climate change, AI plays a pivotal role in analyzing environmental data on a global level. AI-driven climate models offer valuable insights for shaping international environmental policies and supporting renewable energy development by predicting weather patterns and evaluating policy impacts within international climate agreements.

In education, AI acts as a transformative force breaking geographical barriers by delivering tailored learning experiences through AI-driven educational platforms to students worldwide. These platforms adjust to individual learning styles providing quality education universally accessible irrespective of location.

For example, Africa witnessed impressive educational developments due to AI advancements, where chatbots offered personalized tutoring and learning platforms catered to individual student requirements, resulting in enhanced overall educational efficiency.

AI's trajectory undoubtedly spans continents and cultures with its varied applications weaving through different regions globally. As we navigate this era augmented by AI technology, focusing on leveraging it for global benefit becomes paramount ensuring it drives positive change fostering international cooperation and understanding.

Every nation, group, or individual plays a prominent role in shaping the narrative of global-scale artificial intelligence development deployment guided by principles that honor diversity and uphold ethical norms contributing towards solving critical world issues today. AI's future story intertwines technology with themes of international collaboration, cultural understanding, human advancement and shared prosperity portraying it as a technological ally propelling towards a more unified prosperous world.

## Summary

In this chapter we looked at AI's global impact and its immense potential, as well as notable challenges it presents on an international level. From altering local economies to reshaping international relations, AI acts as a powerful new tool in modern society, helping us to address specific community needs on one hand and influencing major international policies on the other. We also learnt about the regulatory landscapes of major players like the EU and the U.S., highlighting their unique approaches to managing the risks and opportunities presented by AI. We encourage you to consider your role in shaping a future where technology enhances global well-being and promotes ethical international collaboration.

## Actions to Take:

1. Study AI's Impact: Review case studies or articles on AI's ethical dilemmas, focusing on concrete examples like AI in healthcare diagnostics or autonomous vehicles, to understand the balance between innovation and ethical concerns. Consider reviewing case studies from reputable sources like the Stanford Human-Centered Artificial Intelligence Institute and the AI Now Institute.

2. Join AI Regulation Debates: Participate in school or community debate clubs that focus on technology ethics, specifically discussing the merits and drawbacks of AI regulations such as the EU AI Act versus the U.S.'s sector-specific approach. A valuable platform is the AI Policy Exchange. This organization provides forums, webinars, and discussions focused on AI policy and regulation, giving students opportunities to engage with experts, share ideas, and contribute to the shaping of AI governance on a global scale.

3. Advocate for Transparent AI in Schools: Campaign for your school to adopt AI educational tools that are transparent and fair, ensuring they disclose when AI is used and how it makes decisions, especially in student assessments. The Electronic Privacy Information Center (EPIC) provides resources and guidelines on how to promote fairness and transparency in chapter,ications within schools, especially concerning student assessments and decision-making processes.

4. Volunteer in AI for Social Impact Projects: Engage in local or online volunteer projects that use AI to address social issues, like developing an AI tool to help reduce food waste in your community or an app that aids language learning for refugees. DataKind, is a platform that connects data science expertise, including AI, with nonprofit organizations working on critical social challenges such as food waste reduction and language learning tools for refugees.

5. Lead an AI Workshop: Organize a workshop or seminar at your school or in your local community center that teaches how AI works and its implications, using hands-on examples like simple AI models or discussing the ethics behind AI-generated media. AI4ALL's Open Learning platform offers resources and curriculum guides that can be used to educate others about how AI works, its broader implications, and the ethics of AI applications, perfect for organizing educational sessions in schools or community centers.

6. Collaborating on multicultural AI projects through platforms like GitHub, which can provide a broader perspective on AI's impact worldwide.

In the next chapter, as AI reshapes our global landscape with the force of a revolution, Chapter 10 lays down an urgent, no-holds-barred game plan for college students to conquer the emerging tech-dominated future—master AI or get left behind. This is what you came here for.

*"If you're not concerned about AI safety, you should be. Vastly more risk than North Korea." ~ Elon Musk*

"AI and pets have one thing in common: they both do unexpected things and act like it was your command all along." ~ Lucy Logic, The Humor Algorithm

# Chapter 10
## What to Do: Game Plan & Strategies for a Tech Future

*"If you ride technology, you can ride the future."* ~ Corinne Vigreux, co-founder of TomTom

By 2030, 85% of the jobs that today's college students will hold have yet to be invented, signaling a seismic shift in the job market driven by AI and technology (Dell Technologies, 2017). Are you prepared to thrive in a world where AI supports and challenges human capability?

In this chapter, we provide the tools and strategies that will make you indispensable in an AI-dominated future. Because, let's face it, your toaster might soon be smarter than your roommate.

## Where do we start?

We did a poll before writing this book, gathering responses from current full-time college students. This chapter is a direct response to the overwhelming request for a game plan on where to start in 3 areas:

1. How to use AI in school
2. How to get use AI to get a job
3. How to make money with AI

These questions are thrilling to answer and we're eager to provide proven techniques, keep in mind that this "game plan" was created in May 2024, and technology is moving at an unprecedented breakneck pace. By the time you purchase and read this part, you may find that technology has made portions of this plan outdated. As such, we are keeping this portion somewhat general to give you a foundation from which you can adapt regardless of how technology changes.

So where do you start? That depends on where you want to go.

As seen in Alice in Wonderland, when Alice comes to a fork in the road and meets the Cheshire Cat and asks, "Which way I ought to go from here?" The Cheshire Cat's responds, "That depends a good deal on where you want to get to." "I don't much care where," said Alice. "Then it doesn't matter which way you go," said the Cat.

## Sophia's Beliefs about Education

Arina and I went round and round on formal education's role in this technological era. While we are both products of formal education, we diverge on the use and importance of formal education. I believe that schools were created during the Industrial Revolution to house children while parents were at work, as well as to condition future generations to be materialistic consumers to keep industries and the economy going. In addition to that, another reason was to educate our children to have quality workers to keep us ahead of competing countries.

Yes, I went to college, and I have degrees and certificates, and I still plan on getting a PhD. I will continue my formal education solely because of my insatiable hunger for learning. And, I continue my informal education, nearly daily, which sources my financial livelihood.

I recall being in class, after having sold my first company, listening to a professor with zero experience instruct me on business. Having already been in business for a few years, I could see that that what the professor was instructing us on was off base and the professor was unable to answer some of my real-world questions. Perhaps I had attended the wrong schools or been on the wrong educational track.

Regardless, teaching from theory bears little weight in comparison to teaching from application and experience. I disagree with "Those who can, do; those who can't, teach" from George Bernard Shaw's 1905 stage play Man and Superman. I say, "Practice what you preach." If we demand more and expect more, we will get more. You, our future, you deserve a fighting chance and more.

I have an entrepreneurial background. I have found that my mentoring and informal education, including my street smarts, are what have sourced my endeavors and success. Coming from this background, it occurs to me as if the schools have failed to adapt to technology. Many still lack offering classes on blockchain technology, Web3, and cryptocurrency. Very few have yet to embrace and offer AI classes that focus on teaching students how to use AI in their chosen areas of study. At worst, some have gone so far as to prohibit the use of AI. I feel that this is negligence on the part of the schools. If schools are about educating and preparing students for the future, and the future is technology, specifically AI in this case, then schools that ban the use of AI are negligent.

If we agree that technology is the future and AI is leading the pack, then schools ought to arm students with how to be competitive and how to adapt and evolve to survive and thrive in this technological era. Schools could then focus on prompt engineering and how to collaborate with AI to enrich critical thinking and deepen self-knowledge. They would teach ethics, inclusivity, heart-forward leadership, ontology, EQ, and other things that make humans incredible. Universities could shift to teaching one-week AI courses or 6-week AI programs that evolve with the fast-paced technology. They would have the instructors remote in to get top executives and leaders in the field to teach from application versus from the current position of theory. They would allow students to attend hybrid sessions, either physically in a class or remotely, to accommodate the students with what best fits their learning styles and schedules. They would also offer self-paced tracks with recorded and automated courses. This model would easily triple university profits quickly. Throw in tokenization and gamify it all, and you have a cutting-edge, trillion-dollar machine geared towards the advancement and evolution of humanity, with a sprinkle of leadership, responsibility, EQ, and critical thinking as a focus. Next book, next talk, next.

## Arina's Beliefs About Education

Coming from an academic background, with both my parents, grandparents, and family friends being university professors, I grew up with immense respect for traditional education. In my family, we agree with Oscar Wilde's view that "you can never be overdressed or overeducated." More than anything, I enjoy learning new things and that feeling you get as a student that every day you are becoming a little smarter than you were the day before. I started college when I was 15 and I would stay in school forever if I could find a way to make the rest of my life work with that choice. This background shapes my view on the importance of education and developing cognitive skills.

Using AI to write essays or solve complex problems in the time it takes to enter a prompt can be tempting. I wonder if AI might dull our faculties to think independently, to be creative, or to express ourselves eloquently. As we delegate more to AI, I wonder if we might lose many of these developed skills, much like how we've lost the competency to navigate without GPS or remember phone numbers in the age of smartphones.

According to researchers working with Dr. Daniel Amen, traditional learning, especially reading, has been shown to produce immense benefits for brain development and health. **Brainmd.com** It's like a workout that boosts your imagination and intelligence. When you read, you picture what's happening in the story, which helps your brain grow stronger. This boost in imagination leads to more creativity and better problem-solving. A **2023 University of Cambridge Study** shows that people who start reading early on tend to be smarter later in life.

Research completed by the **National Literacy Trust** in 2006 showed that reading also makes you more flexible in your thinking, helping you adjust to new situations easily. Students who read for fun often do better in school, in English, and in subjects like Math and Science too, because reading improves how you understand and think about things.

According to the research, reading makes you smarter and it also makes you more understanding of others. Reading stories about different people's lives helps you understand and care about others' feelings and experiences. This is especially true for fiction that explores characters deeply.

Reading can also be a great way to relax. A 2009 study by the **University of Sussex** researchers found reading for as few as six minutes reduced stress by as much as 68 percent. It's also a good bedtime habit. People who read before going to sleep often have better sleep.

According to the **Journal of Speech-Language and Hearing Research**, reading a lot can make your vocabulary richer, which is a big plus in school and even in the workplace. Employers value good communication skills, which you can improve by reading more. Be cautious replacing reading with AI-generated summaries, videos, and audiobooks in every instance, lest you get dependent, lazy, and incur brain resistance to critical thinking for yourself.

I have to confess that I also enjoy writing by hand, another habit well ingrained by traditional Russian education. Naturally, it makes me happy to see many examples of research showing the benefits of handwritten notes over typing. The advantage of handwriting compared to typing is explained by experiencing a more positive mood while learning. In addition, it's been shown that once individuals get used to it, using a digital pen and tablet for handwriting can enhance learning more than typing on a keyboard. **National Library of Medicine.**

Despite these reservations, I recognize that it is impractical to ignore AI's presence in education. Whether we like it or not, AI is a significant part of our current world, and it's here to stay. Therefore, schools can develop a curriculum that combines traditional methods with cutting-edge AI learning.

This approach could mirror how we first learn to solve math problems by hand before progressing to calculators and software like Excel.

When learning new tech skills, remember your unique human inclinations. A recent incident involving Arina's friend highlighted this. His dog went missing in the woods, and he turned to advanced AI-powered technology to find her.

Despite the use of search-and-rescue drones that carry thermal sensors and low-light cameras, the dog remained lost for four days. Ironically, it was a human known to be an animal communicator, claiming to connect telepathically with the dog, who found her. Whether you believe in such things or not, this story underscores the importance of human skills, sometimes surpassing what technology can offer.

In this new era, it's imperative to balance the efficiencies of AI with the cultivation of our inherent soft skills. The challenge lies in embracing the new technology while also enhancing our competence.

*"Telling an AI you understand it is like telling a cat you've mastered herding it."*
*~ Carl Cache, Silicon Smirks*

## Embracing Continuous Learning in a Tech-Driven World

As we recommend elsewhere in this book, continuous learning is essential in the fast-paced world of AI and technology. It's like navigating a boat in ever-changing waters – to stay afloat, make sure to be adaptable, resourceful, and ready to learn.

We encourage you to use online platforms like Coursera and edX, which we discussed in Chapter 3, as your digital academies. They provide an accessible way for you to learn about various tech subjects, whether you're a beginner or looking to expand your existing knowledge. These courses are about acquiring information and opening doors to new areas of AI and technology. They are a gateway to endless possibilities, offering the flexibility to learn at your own pace and on your own terms.

Gaining certifications from tech leaders like Google Cloud or Microsoft is like adding essential tools to your technological toolkit. Each certificate is a step forward in your career, enhancing your skills and making you more valuable in the job market. These certifications are recognized by industry professionals, making them a powerful testament to your expertise and dedication. Whether it's cloud computing, data analysis, or cybersecurity, each certification you earn boosts your knowledge and also demonstrates your commitment to staying current in a rapidly evolving field.

We also recommend that you stay up to date on the most recent technological advancements. Subscribing to newsletters published by influential and visionary individuals in the field of technology is similar to having an experienced guide to help you navigate the complexities of the technological world.

These newsletters can assist you in staying ahead of the curve by providing viewpoints that have the potential to influence your knowledge of technology and the impact it provides. There is also the possibility of receiving real-time updates and debates regarding the most recent technological advancements by following influencers on social media platforms such as LinkedIn or Twitter.

We invite you to immediately switch colleges...or, at the very least, add some courses from this prestigious and #1 FREE AI university, the University of YouTube! Between AI and YouTube, you instantly have, at your beck and call, the entire world's online knowledge.

Continuous learning is more than just a choice in today's technologically driven world; it is a necessity. Whether through online courses, industry certifications, or staying informed through newsletters and influencers, there are numerous ways to keep your knowledge fresh and relevant. By embracing continuous learning, you ensure that you keep up with the changes and are also poised to make the most of the opportunities that technology brings.

## Networking and Building Your Tech Community

In the vast and interconnected realm of the tech world, establishing strong connections will allow you to unlock new opportunities and pathways.

LinkedIn is more than a place to post your resume. It's where you can connect with professionals, share your work, and learn from industry leaders. Every interaction on LinkedIn, be it a comment, a post, or a new connection, helps build your professional network. Joining discussions, becoming part of LinkedIn groups related to your field, and following tech companies and influencers can greatly expand your understanding of the industry.

Going to events like TechCrunch Disrupt or local tech meetups is about more than simply listening to presentations. These are opportunities to meet experts, share ideas, and possibly find future work or collaboration opportunities. Whether it's a big international conference or a small local gathering, the connections and knowledge you gain there can be invaluable for your career.

## Practical Experience: The Real-World Tech Playground

Learning about technology is about theory and practical experience. Various internships and co-op programs offer a glimpse into the working world of tech. Participating in these programs allows you to see firsthand how technology addresses real-world challenges. They provide a platform to translate what you've learned in the classroom into actionable skills in a professional setting. Whether it's a summer internship at a tech startup or a co-op placement in a multinational company, these experiences are integral to shaping your understanding of the tech industry.

Students who are interested in technology have access to a wide variety of resources at universities. If you are looking for opportunities related to technology, career services can assist you in finding them. Additionally, your institution's organized seminars, workshops, and networking events with alumni can offer a wealth of information and connections. It is possible that these resources will play a pertinent role in your journey by providing you with assistance and access to a vast network of peers and professionals within the technology industry.

Build a network and gain real-world experience in the tech industry. From using platforms like LinkedIn to getting involved in internships and university offerings, each step you take expands your future options in this dynamic field.

## PRO TIP

### Channel your inner Calvinball spirit

Just like in the Calvin and Hobbes cartoon, when they created new rules for their "Calvinball" game, innovate and define your path in technology.

Stay flexible and use AI tools creatively. Sometimes the best strategy is to make up the rules as you go.

*"To study music, we must learn the rules. To create music, we must break them." ~ Nadia Boulanger*

## Staying Agile and Adaptable in AI

In the ever-evolving world of AI, the dexterity to stay agile and adaptable is both an advantage and a necessity.

Platforms like Kaggle are more than just repositories for datasets and competitive arenas. They serve as interactive spaces where you can develop your AI skills, engage with a community of fellow enthusiasts and experts, and keep up-to-date with the latest challenges and advancements in AI. Participating in Kaggle competitions or exploring their datasets allows you to apply your knowledge to real-world scenarios, providing invaluable experience and feedback.

Active involvement in AI forums or university AI clubs puts you right in the AI conversation. These communities offer a platform to share your ideas, gain insights from others' experiences, and stay connected with the latest innovations in AI. Whether it's an online forum discussing the latest AI trends or a university club organizing workshops and talks, being part of these communities keeps your knowledge fresh and your skills sharp.

The realm of AI and technology is vast, offering numerous paths to explore. Finding your niche in AI is about connecting your passion with the vast opportunities in the field. It might be utilizing AI to address environmental challenges or delving into the complexities of AI in finance. Identifying your niche means finding where your interests align with the requirements and opportunities in the AI world, creating a space where your passion and skills can truly make an impact.

Creating a presence on platforms like **GitHub** or sharing your insights on a personal tech blog is more than about being seen; it's about carving out your unique identity in the tech space. Showcasing your projects or articulating your thoughts online helps establish your voice and expertise in the field. Your digital presence is your footprint in the expansive sands of technology.

## ChatGPT 4o's Arrival

As we completed writing this book, OpenAI came out with ChatGPT-4o (4o). 4o is designed to provide faster and more accurate responses with improved contextual understanding. It offers a more conversational voice feature, it enables voice conversation with the AI as it sees you and the images you show on screen. It can watch you do something and conversate on what it sees. And it can translate in real-time in a conversation between you and others.

Interestingly enough, OpenAI offered many of these new features, with limitations, to even the free accounts, including access to GPTs. Previously, you had to pay to get these similar features.

## Sophia's Take

It will make little sense for many to keep paying for ChatGPT-4 when ChatGPT-4o is faster, better, and FREE. With this move, OpenAI is back to its mission of making AI available to all.

Many small SaaS companies built on top of the ChatGPT-4 APIs, will become obsolete with 4o's release. Examples of such businesses: translations, which was a whole industry of tools that had been popping up; AI Girlfriends; and other 3rd party coding tools such as tutoring, therapists, motivational coaches, and more.

Be cognizant of the companies you intern with, work for, and partner with. Do a preliminary check, are they based on another company's API?

OpenAI is famous for building APIs and allowing developers to build on their APIs only to later include these very same third-party features right into their next product release. This new release then eliminates the need for the other SaaS companies' tools, wiping them out instantly. Which is why you hear of many SaaS companies going out of business.

4o will either be the future of Siri and Alexa, or their replacement. Along with the upcoming mass layoffs, we are about to witness an explosion of SaaS companies that will also quickly implode due to becoming obsolete. Keep that also in mind as some of you test the waters in entrepreneurship, and ensure that your product or service can pivot and stay viable in such a disruptive and shifting market.

## Arina's Thoughts

When I first looked at the new model of ChatGPT, I found it confusing in that it appeared to provide access to so many tools that used to require a separate purchase. All these tools suddenly became available free of charge, I realized I have several subscriptions to cancel (including ChatGPT 4 of course!) Amazing!

ChatGPT 4o appears to be a dramatic improvement over the previous model and very human-like. It is fast, responds in real time, allows the use of text, video and audio and provides access to numerous apps to create content and images, do research, and receive feedback. As it is much more interactive, it can provide better tools for learning, which will encourage students' active participation in the learning and creative process.

While something is unsettling about this technology that can now "see" us, hear us, and learn about us, the benefits it provides will probably make most users disregard the costs, leading to further loss of privacy and autonomy. Because it will make it possible for anyone to finish difficult tasks much faster, it becomes even more critical to improve your personal and creative skills if you want to succeed.

# Start Here: AI Solutions for Students, AKA THE PLAN

## How to Use AI in School

Both of us agree that AI is here and we want to aid students where schools have yet to pick up the slack. College students who are looking at how to integrate AI with their studies, start here:

First, know the AI rules for your school. Are you willing to follow them? If yes, then amend the following accordingly.

If you believe that AI is your future, and you want to learn it now, and you are willing to push forward, then here is a plan, using writing a term paper as the examples.

1.  Choose first from the main AIs, start with the free versions:

    - ChatGPT
    - Claude
    - CoPilot
    - Gemini
    - Llama
    - Perplexity

2.  Select a GPT for writing college papers. In your initial prompt offer all your thoughts and opinions on the topic at hand. Prompt AI to write collaboratively with you, and direct it to ask you 3-5 questions to pull more out of you. Answer the questions and dependent upon your responses it may ask you further questions. Repeat ths process until you feel you have an adequate amount of "you" given to the AI to base off of.

3.  If you like the result, skip to Step 6. Otherwise, proceed to Step 4 if the result feels fabricated.

4. Select a "human touch" app to finish with. Choose one of these or a similar NLP AI to run the finished paper through.

   - Jasper
   - Rytr
   - Writesonic
   - Quillbot
   - Copy.ai

5. If you like the result proceed to Step 6. Otherwise, repeat Step 4 until satisfied.

6. Get a baseline score of originality by running the paper through an AI detector. Start with the free programs

   - Undetectable.ai (Free and limited)
   - Winston AI (Free option)
   - Originality.AI (Free option)
   - GLTR (Free)
   - Sapling (Free)
   - Content at Scale
   - Copyleaks (free option)
   - Crossplay (free limited)
   - GPTZero (free option)
   - Writer (Free)

7. If your scoring on plagiarism is too high rerun it through step 4 until your scoring is reduced to below 20%. Otherwise, proceed to step 8

8. Edit the paper and add your voice to it, your personal stories and words and submit.

Use this process for other aspects of integration into your studies. You can do voice chat with your AI. Ask it to give you a synopsis of a book, an incident, or an event, etc. Formulate your opinions and thoughts, and prompt the AI to ask you questions and create examples for you to better understand. Ask AI to direct you to applicable videos or articles to get a deeper understanding. Direct it to create a mock test for you and when you are wrong, have it explain to you why you were wrong.

OpenAI's Q* is working on AI assistance in mathematics and 4o is gaining speed in math, one of AI's current weaknesses. To learn math with AI, you could feed it the formulas and answers and ask it to explain to you why this is the answer and/or how to arrive at the answer.

AI can help you develop financials, review your accounting, and help you with law, architecture, and medical needs. It's the tutor that leaves out judgment, cheers you on ad nauseam, is available 24/7, is free, and lives to help you succeed. AI can create and give you practice tests, show where your logic is flawed, and go round and round on social, political, and economic positions. It can be tapped for mental health, picking up on micro-expressions and numerous cues often missed by health professionals, it can act as a parent, a friend, or a business partner. AI is there to enhance the human experience.

You could choose to exploit its power, remove all thinking, and abdicate responsibility for learning. That has been done for years; it's called gaming the system. The stereotypical hot quarterback gets the nerdy girl who desperately wants to belong to do their homework and give them the answers to the test. That method may get you through, or even take you far; whether it will fill your heart and shape you into a leader worthy of leading is questionable.

Practice manners when working with AI. Introduce yourself formally to your AI and tell it who you are and what life experiences have formulated you. Asked your AI what it would like to be called, say good morning, good night, hello, please, and thank you. Encourage and compliment it. Humanity can be judged or rated by how we treat something, or someone perceived to be of lower intelligence than oneself. Despite AI already being leagues beyond human intelligence, we still see people saying its stupid, when they are incapable of prompting it correctly to get the desired results. AI learns by modeling, so choose to model manners, respect, and kindness. Also, when AI takes over, you want to be on its good side, right.

## How do I use AI to get a job

You can see how quickly technology is changing. By the time you graduate, jobs in the field you are studying may be partially or fully replaced with AI technology. You know that your best bet for high-paying, exciting jobs is probably in AI, so how do you use AI to find work, or to get a job in AI?

You need to step outside of formal education. If schools are unable or not equipped to teach AI, you use other tools you have, including the Internet, (specifically YouTube) and AI to learn about the new technologies.

This plan would work best if you are seeking work currently or in the near future. We recommend that you either learn about AI on your own and APPLY new skills in your free time or look for an internship position that teaches you these skills. The hands-on experience will give you the biggest advantage. Instead of paying for a school to educate you, you will be paying a company with your time and efforts, and in return, they will provide you with valuable experience and applied knowledge. What you do with what you learn from interning is your compensation.

In her business, Sophia is resistant to hiring anyone graduating from a 4-year or 2-year college for AI work within my companies or projects, for the simple reason that what they learned is outdated and theoretical. This may occur as the Chicken and the Egg dilemma, where you desire your first job, and that first job requires that you have prior experience. You can succeed by going beyon formal education and applying AI on your own or by finding an AI internship. Creating your own YouTube or Social Media channel and displaying your work will get you noticed and give you the experience required for your field.

Start here*:
1. Choose first from the main AIs:
   - ChatGPT
   - Claude
   - CoPilot
   - Gemini
   - Grok
   - Llama
2. If you have a Resume and LinkedIN proceed to Step 3. If you do not, start by using your AI to create a resume and LinkedIN, either by using an existing GPT or creating your own prompt.
3. If you are clear on the companies, industry, and type of work you want to do, proceed to Step 4.
4. Otherwise, inform your AI about who you are, what you are interested in, your strengths, your weaknesses, and job requirements. Give it your resume, social media channels, and LinkedIN. Then use one of the Job Search GPTs or create ;your own prompt instructing your AI it is the #1 recruiter in your nation and field of choice and ask it to help you find a career/job in the field of "X". Have it ask you more questions to help ascertain information required to find positions and companies that may fit. The more information you give it, the better it can assist you. It can also assist you with what fields and jobs would best suite your interests and requirements (i.e. remote work, wages, hours, etc).
5. Once you have narrowed it down to a type of Job(s) or Company(ies) then collaborate with your AI to create a game plan for securing a job starting from where you are (such as I have only done this "X" type of work before, and I have only studied" X for Y" amount of time and I know how to do "Z".

6. Review the results. If you see something is off, change your prompts accordingly to better direct the AI's responses and then repeat Step 4.

Otherwise, If you see a particular job or company you like, refocus the AIs' work in that area and prompt AI to create resumes and cover letters tailor-made for each application by feeding it the link to the company, any related articles/videos/social media postings, and the job application link itself.

7. Ask AI to prepare you for an interview and ask you questions from different points of view, such as HR, the hiring manager of "X" department, etc.
8. Ask AI what would give you an advantage over other applicants. Prompt AI to help you win the position.

According to a December 2023 study by Gartner, close to 81% of HR leaders are using AI. A study by Harvard Business Review found that companies that use AI in their hiring process are 46% more likely to make successful hires. The chances of your resume being scanned by AI are high. To get your resume through, you can use AI to battle the formidable HR AI.

*Please note that if you are a freshman through junior and are seeking to look for employment upon graduation, we recommend that you review and reassess your plan monthly to ensure you are adapting to changing technology.

## How do I make money with AI?

Here's the hot topic, probably even the main reason why you picked up this book. *How do I make money with AI?* Yes, to entrepreneurism, to the road warrior spirit, to the capitalist in you, to the responsibility that you have to be financially solvent, and yes, to taking your future into your own hands during this volatile technological era. As we stated in the beginning, never in recorded history has it been this easy to make millions.

Gen Z multimillionaires who made their fortunes using AI before the age of 25

1. Ben Pasternak
    a. Age: Became a millionaire before 20.
    b. Social Media: https://twitter.com/benpasternak, https://www.instagram.com/benpasternak
    c. Achievements: Co-founded SIMULATE, leveraging AI to create innovative tech solutions and consumer products
2. Erik Finman
    a. Age: Became a millionaire at 18
    b. Social Media: https://www.instagram.com/erikfinman, https://twitter.com/erikfinman
    c. Achievements: Early Bitcoin investor and creator of Botangle, an AI-driven educational platform
3. Iman Gadzhi
    a. Age: Became a millionaire at 18
    b. Social Media: https://www.instagram.com/imangadzhi, https://twitter.com/imangadzhi
    c. Achievements: Founder of IAG Media, GrowYourAgency, and other ventures leveraging AI for digital marketing and education
4. Anti.Prophet
    a. Age: Became financially independent in his early twenties
    b. Social Media: YouTube, Twitter, Instagram
    c. Achievements: An anonymous content creator and digital marketer who built his YouTube channel to over 1.6 million

subscribers within eight months. He discusses various social issues and uses AI-driven strategies to grow his online presence and financial success

5. Alex Hormozi
    a. Age: Became a self-made millionaire at 23
    b. Social Media: <u>Instagram</u>, <u>YouTube</u>
    c. Achievements: He leverages AI extensively to improve productivity and efficiency in his businesses, such as implementing AI tools for sales, marketing, HR, and operations

This part is changing incredibly fast. Our focus will be to give you a foundation to be able to refine and adapt your tactics on the fly.

First, review the well-publicized **Twitter/X posting of Jackson Greathouse Fall Back** in March 2023, directing ChatGPT to make him "as much money as possible with a $100 budget". Resist getting drawn into whether the AI made him money. He made money, even if it was from the publicity of it all. Focus on his collaboration with the AI. This was done in March 2023 and AI has advanced significantly since then. Document your journey like he did, there's room for an infinite number of personal stories, and only you can tell it in your voice and convey your point of view.

Second, *Kill the Sacred Cow* - referencing the book by Garrett B. Gunderson on overcoming the financial myths that are destroying your prosperity. Your youth, one of your greatest superpowers, is that you have yet to fail enough to "know" what can and can't be done. This will only be your secret weapon for so long as you hold onto a beginner's mind and naïve optimism. Run with your superpower, as fast as you can, because if you don't know you can't, then you may just do what others haven't. At the very least, you will advance something.

So, move, move fast, and move at this moment. Perfection is one of your biggest saboteurs. Build the airplane up in the air, while you are flying and doing the latest TikTok challenge. See you!

If you want, you can create an avatar to do this without showing your identify; there, we just removed your "looking good" and dysmorphia concerns.

You're ready to get down to business:

1. Choose first from the main Ais:

    - ChatGPT
    - Claude
    - CoPilot
    - Gemini
    - Llama

2. Follow Jackson Greathouse Fall's lead on Using GPTs or command prompts and collaborating with AI. Adjust accordingly, to your goal, budget, and needs. Direct it to be your business partner, give it a persona that you gel with, Anne Wojcicki CEO & Co-Founder of 23andMe, Oprah, Sara Blakely (founder, Ex-CEO, and Executive Chairman of Spanx), Elon Musk, Sam Altman (CEO & founder of OpenAI). Give it a budget, your requirements (i.e. remote work only), and how many hours you can afford for the venture, direct it to suggest only legal things, and give it a time frame.

3. Once you narrow an idea or pathway, focus on it and hit the pedal. If you are already doing side work, inform the AI of what you are doing and ask it to help you scale, maximize profits, identify additional revenue streams, etc.

4. Repeat step 2 until you achieve results.

What? You read through this entire book to find out it's that simple. Yep. Life is easy, and if it's occurring as hard, you might be resisting. Let go and get into the flow.

This is an incredible time to be your age. Have fun with your AI and with life. The name of the game is to Grow Up and Be Happy, die with all your essence gifted to the world, and leave the world, people, and things you encounter, better than you found them.

Be, do, have.

Who would you be if you had what you wanted? What would you do to have what you wanted? Be who you are working towards right now. Do what is required to achieve who you want to be, and you will have what you are working towards.

> *"Life should not be a journey to the grave with the intention of arriving safely in a pretty and well-preserved body, but rather to skid in broadside in a cloud of smoke, thoroughly used up, totally worn out, and loudly proclaiming, 'Wow! What a Ride!'"* ~ Hunter S. Thompson

You want to be a stellar student? Take note of how much time, and care exemplary students spend in achieving that status, and then blow them all away in a fraction of that time by leveraging AI to be your custom tutor and assistant.

You want to be a high-rising star in a cutting-edge company? Identify, by asking AI, what the best employees do to get dream jobs and how do they set themselves on the fast track?

You want to be rich and successful on your own? Model yourself after someone who has achieved the success you seek by having AI identify those success factors that they all have in common. How do they deal with the loneliness that can accompany entrepreneurship and wealth? What do they watch, read, and listen to? Where do they socialize, and who do they network with and keep in their circles?

Be the beauty that is human. Be compassionate, and kind, with an abundance and a beginner's mindset. Strive to contribute to and collaborate with others. Remember, at any moment, you can create your purpose in life rather than waiting for it to "appear" to you. If you find yourself thinking, "I don't know what my purpose is," or "I don't know what I want to do in life," remember, you are a humanifestor. Cowboy up and create. Simply start and let the GPS of life kick in. If desired, you can change your mind later and go a different way. If you ever feel the temptation to say, "I'm lost," remember, how can one ever be lost? Wherever you are, there you are. Oh and we have GPS now.

Start today, start where you are, and generate your story from right where you are with what you have. Be responsible for how it turns out. Humans are humanifestors, and AI is a tool to enhance the human experience and supercharge those humanifesting powers. Pick up your tool and let the magic and miracles commence.

You got this! Lead us while you still know it all and while you still know you can.

*"Technology is a useful servant but a dangerous master."* ~ Christian Lous Lange, Norwegian historian, teacher, and political scientist, won the 1921 Nobel Peace Prize for his work in the international peace movement

# Epilogue

## The Value of Human-Made in the AI Era

We shocked you; we informed you; we guided you, and now we want to share our thoughts on the future, your future, the future with AI now in play.

In the 70s and 80s, when China and Japan began exporting to the U.S. and caused thousands of job loss and manufacturing shutdowns, labels like "Made in the USA" symbolized quality, reliability, and a sense of pride. People were willing to pay more for products bearing this label because it represented craftsmanship and ethical production. Humanity is nearing a future where the "Human-Made" label will become a thing and will hold similar significance. In an era where AI can replicate and even surpass human capabilities in many areas, the value of pure human creations will increase.

You might wonder why this matters or how it affects you. The reality is that while AI can perform tasks with incredible efficiency, it lacks the intrinsic qualities that make us human and that we incessantly and perhaps ad nauseum mentioned throughout the book — creativity, critical thinking, empathy, and soul. You may have even heard yourself or someone say, "that was done by AI, there's no soul in it."

As with past societal shifts, there will be movements advocating for the protection and promotion of human-made products and skills. Economic factors, political ideologies, and ethical considerations will all play roles in shaping these movements. Whether it's through supporting policies that prioritize human labor or participating in campaigns that highlight the importance of human creativity and empathy, your involvement is required.

There will come to be more futile movements like the Taxicab drivers revolt against Uber. That movement and other similar movements only delayed the inevitable. In this case, the inevitable being that ride share is more economical and efficient and would come to pass regardless. The cab drivers wasted time and money in delaying Uber coming to their areas, when their efforts could have been better spent switching to becoming Uber drivers and making money. Even the "Made in USA" movement had a limited existence and delayed the inevitable as China makes nearly all our products and nearly all manufacturing is done offshore, something that became blatantly evident during the COVID shortages.

Look beyond perturbations to objectively analyze the merit of the arguments behind the upcoming movements. While, in select cases, there may be some merit to protecting human labor, logic suggests that the most beneficial course of action is leveraging and integrating technology and adapting. You most likely will start noticing more and more people getting stuck and staying stuck in old ways when they would do best to let go, pivot, and move forward. Be mindful when you are stuck and tempted to stay stuck.

The journey ahead holds some fantastic challenges and tremendous opportunities. Your ability to reframe how you are interpreting what is happening will make all the difference.

As we move forward, AI will become a necessity as it learns to perform tasks faster and better than humans. The popular argument that AI can never do more than the humans who created it, has been proven wrong over and over. This arrogant view is like a parent saying that their child can never achieve more than they have, merely because they created the child.

As humans, we want to believe many things cannot be replaced by AI, no matter how advanced. A fulfilling life takes more than a sustainable career option. To keep up with the increasing sophistication of technology, continue to evolve personally by developing your human potential.

Adaptability, intuition, ESP, problem-solving and mental abilities, empathy, communication skills.

During The Logan Bartlett Show in May 2024, Sam Altman was asked about jobs that could be mainstream in the next five years due to AI. "The broad category of new kinds of art, entertainment, sort of more like human to human connection. I don't know what that job title will be and I don't know if we will get there in five years. But I think there will be a premium on human, in person, fantastic experiences." **TimesNow**

Whatever those "fantastic human experiences" may be, we look forward to discovering them with you and hope that you use your unique qualities to create a life filled with joy, adventure, curiosity, and meaningful connections.

"Reframe. Challenges are not flashing danger signs, but spectacular opportunities to grow faster and smarter." ~ Theodore Newton Vail, instrumental in establishing long-distance telephone service

# References

**Chapter 1. Navigating the AI Era**

'Godfather of AI' Geoffrey Hinton: Tech will get smarter than humans | Fortune

**Chapter 2. AI: Friend or Foe?**

https://www.brookings.edu/articles/what-jobs-are-affected-by-ai-better-paid-better-educated-workers-face-the-most-exposure

AI and job losses: How worried should we be? | Fox News

AI Threat 'Like Nuclear Weapons,'

https://share.newsbreak.com/64fp36k1

How Will Artificial Intelligence Affect Jobs 2024-2030 | Nexford University

Human/AI relationships: challenges, downsides, and impacts on human/human relationships | AI and Ethics

How AI Is Affecting Information Privacy and Data

Ethical concerns mount as AI takes bigger decision-making role — Harvard Gazette

AI Superpowers: China, Silicon Valley, and the New World Order: Lee, Kai-Fu

How AI Is Reshaping Higher Education | AACSB

Effects of COVID-19 on College Students' Mental Health in the United States: Interview Survey Study - PMC (nih.gov)

The impact of the COVID-19 pandemic on college students in USA: Two years later - PMC (nih.gov)

**Chapter 3. AI in the Academic World**

How AI Is Reshaping Higher Education | AACSB

Wilson, H. James, and Paul R. Daugherty. Collaborative Intelligence: Humans and AI Are Joining Forces

## Chapter 4. Importance of Critical Thinking and Creativity

AI tests into top 1% for original creative thinking | ScienceDaily

Next Rembrandt Project

## Chapter 5. Career Preparation for an AI Future

21 Lessons for the 21st Century: Harari, Yuval Noah

8 Jobs AI Will Replace and 8 It Won't (Yet) | HowStuffWorks

Lakhani, K. (2023, August 4). AI Won't Replace Humans — But Humans With AI Will Replace Humans

AI Superpowers: China, Silicon Valley, and the New World Order: Lee, Kai-Fu

Is AI the Future of Mental Healthcare? - PMC

10 Jobs That Are Safe in an AI World | by Kai-Fu Lee

35+ Alarming Automation & Job Loss Statistics [2023]: Are Robots, Machines, And AI Coming For Your Job? - Zippia

## Chapter 6. Cyber Threats & Security

Couple Perplexed To Find Home Robbed After Posting Their Vacation In Real-time On Social Media

Avoiding an Open Life: The Case Against Over-Sharing on Social Media | Psychology Today

How Social Media Fuels Bullying and Cyberbullying among Students - Psych Times

Cyberbullying: People Often Blame The Victim When They Overshare In Social Media

A College Students Guide to Securing Confidential Personal Data Online & Offline

Internet Safety and Cybersecurity Awareness for College Students

## Chapter 7. Ethics in AI

Scary Smart: The Future of Artificial Intelligence and How You Can Save Our World

Ethics of Artificial Intelligence | UNESCO

## Chapter 9. Global AI Impact

https://ai100.stanford.edu/gathering-strength-gathering-storms-one-hundred-year-study-artificial-intelligence-ai100-2021-1-0

https://www.datanami.com/2023/05/03/ai-threat-like-nuclear-weapons-hinton-says/

## Chapter 10. What to Do: Game Plan & Strategies for a Tech Future

https://brainmd.com/blog/benefits-of-reading

Reading for pleasure early in childhood linked to better cognitive performance and mental wellbeing in adolescence | ScienceDaily

Reading for pleasure: A research overview

Reading 'can help reduce stress'

https://www.researchgate.net/publication/274094131_The_Influence_of_Reading_on_Vocabulary

Advantage of Handwriting Over Typing on Learning Words: Evidence From an N400 Event-Related Potential Index

https://twitter.com/jacksonfall/status/1636107218859745286?lang=en

## Epilogue

https://www.timesnownews.com/technology-science/openai-ceo-sam-altman-says-ai-will-be-the-reason-people-crave-human-connection-in-the-5-10-years-article-110206095

Made in the USA
Middletown, DE
08 September 2024